ÉCLAIR

by GARUHARU

GARUHARU MASTER BOOK SERIES 1

ÉCLAIR by GARUHARU

초판 1쇄 발행 2020년 5월 20일
초판 5쇄 발행 2023년 2월 20일

지은이 윤은영
영문번역 김예성
펴낸이 한준희
발행처 (주)아이콕스

기획·편집 박윤선
디자인 김보라
사진 박성영
스타일링 이화영
영업·마케팅 김남권, 조용훈, 문성빈
경영지원 김효선, 이정민

주소 경기도 부천시 조마루로385번길 122 삼보테크노타워 2002호
홈페이지 www.icoxpublish.com
쇼핑몰 www.baek2.kr (백두도서쇼핑몰)
인스타그램 @thetable_book
이메일 thetable_book@naver.com
전화 032) 674-5685
팩스 032) 676-5685

등록 2015년 07월 09일 제 386-251002015000034호
ISBN 979-11-6426-105-5

- 더테이블은 '새로움' 속에 '가치'를 담는 취미 · 실용 · 예술 출판 브랜드입니다.
- 더테이블은 독자 여러분의 도서에 대한 의견과 투고를 기다리고 있습니다.
  책 출간을 원하시는 분은 더테이블 이메일 thetable_book@naver.com으로 간략한 기획안과 성함, 연락처를 보내주세요.

- 잘못된 책은 구입하신 서점에서 바꾸어드립니다.
- 책값은 뒤표지에 있습니다.

GARUHARU MASTER BOOK SERIES 1

ÉCLAIR by GARUHARU

에클레어 바이 가루하루

윤은영 지음

더테이블 THE TABLE

PROLOGUE

열다섯. 작은 손으로 빚어낸 밀가루 반죽이 오븐 속에서 근사한 과자로 부풀어 오르는 모습을 지켜보았던 날, 제 마음도 따뜻하고 달콤한 과자처럼 부풀어 올랐습니다. 울퉁불퉁 오트밀 쿠키 한 조각을 맛보고 엄지를 추켜세우며 한껏 웃던 가족들과 친구들의 행복한 표정이 저를 '파티시에'라는 길로 이끌어 주었습니다.

파티시에는 다른 이들에게 달콤한 맛의 휴식과 즐거움을 주는 행복한 직업이지만, 기술을 배우고 능숙해지기까지 긴 시간과 노력이 필요했습니다. 하지만 이 길고 힘든 시간 끝에는 저와 제가 만든 제품들이 함께 성장해 있었습니다.

단지 레시피가 아닌 그 과정에서 겪었던 많은 시행착오와 실패를 경험하며 터득한 포인트와 팁, 도구 활용법은 물론 제조 공정에서의 잦은 실수를 줄여주는 방법들을 이 책에 담았습니다. 또한 해외 마스터 클래스를 진행하며 만난 다양한 문화권의 훌륭한 셰프들과 새로운 식재료들로부터 받은 영감의 결과물을 오롯이 담고자 노력했습니다.

이 책이 누군가에게 새로운 영감을 주는 도구로 사용되길 바랍니다. 이 책의 레시피를 토대로 여러분의 주변에서 구할 수 있는 재료를 활용해 다양한 시도를 해보셨으면 좋겠습니다. 제가 늘 작업대 위에서 기존의 제품과 새로운 재료의 조합을 고민하듯, 여러분의 작업실에서도 이러한 고민과 새로운 시도가 계속되길 희망하며, 그 과정에서 저의 책이 작은 도움이 되었으면 합니다.

끝으로 이 책을 펴내는 데 도움을 주신 많은 분들과 아낌없는 지원을 해주신 더테이블 관계자 분들께 고마운 마음을 전합니다.

윤은영

Fifteen. The day I watched the dough I made with my small hands rise up into gorgeous cookies, my heart billowed just like the warm and sweet cookie. The happy faces of family and friends who tasted a piece of the lumpy oatmeal cookie, raising their thumbs up with big loving laughers, led me to the path of becoming a 'pâtissier.'

Being a pâtissier a blissful job, giving others a sweet taste of relaxation and delightfulness, but it took a long time and effort to learn the skills and become proficient. However, at the end of this long and challenging time, I and the products I made have been growing together.

This book is not just a collection of recipes, but also reflects a lot of trials and errors that we have experienced along the way, and tried to cover the points and tips we've learned, including how to utilize tools, as well as ways to minimize frequent mistakes during the process of making. I also tried to capture the results of inspirations from the talented chefs we've met from various cultures and new ingredients we came across while conducting the overseas master classes.

I hope this book will serve as a new inspiration for someone. Please make a variety of attempts with ingredients available near you by utilizing the recipes in this book. Just as I always contemplate the combination of existing products and new ingredients on my workstation, I hope these thoughts and further attempts will continue in your studio and wish my book will be of little help along the way.

And last but not least, I would like to thank many people who helped me publish this book and the officials of THETABLE publishers for their generous support.

Yun Eunyoung

HOW TO USE THIS BOOK

이 책을 활용하는 법

데크 오븐과 컨벡션 오븐 사용법을 알려드립니다.
데크 오븐은 윗불과 아랫불을 각각 조절해야 하며 제시된 데크 오븐 아이콘은 '윗불 190℃/아랫불 190℃에서 5분간 굽다가 아랫불을 170℃로 내려 50분간 굽는다.'는 뜻입니다. 제시된 컨벡션 오븐 아이콘은 '160℃ 오븐에서 30분간 굽는다.'는 뜻입니다. 데크 오븐과 컨벡션 오븐에 대한 설명은 본책 35p를 참고하시기 바랍니다.

Instruction for deck oven and convection oven are as follows; For the deck oven, the top and bottom heat should be adjusted separately. The proposed icon means, 'bake at upper heat 190°C / lower heat 190°C for 5 minutes, then reduce bottom heat to 170°C for 50 minutes.' The proposed icon for convection oven means 'bake at 160°C for 30 minutes.' For explanations on the convection oven and deck oven, please refer to page 35 in this book.

슈 반죽 레시피는 12cm 에클레어 기준 56개가 완성되는 배합입니다. 필요한 만큼 배합을 줄이거나 늘려 사용할 수 있습니다.

The choux recipe yields 56 pieces of 12cm éclair. The recipe can be divided or multiplied as needed.

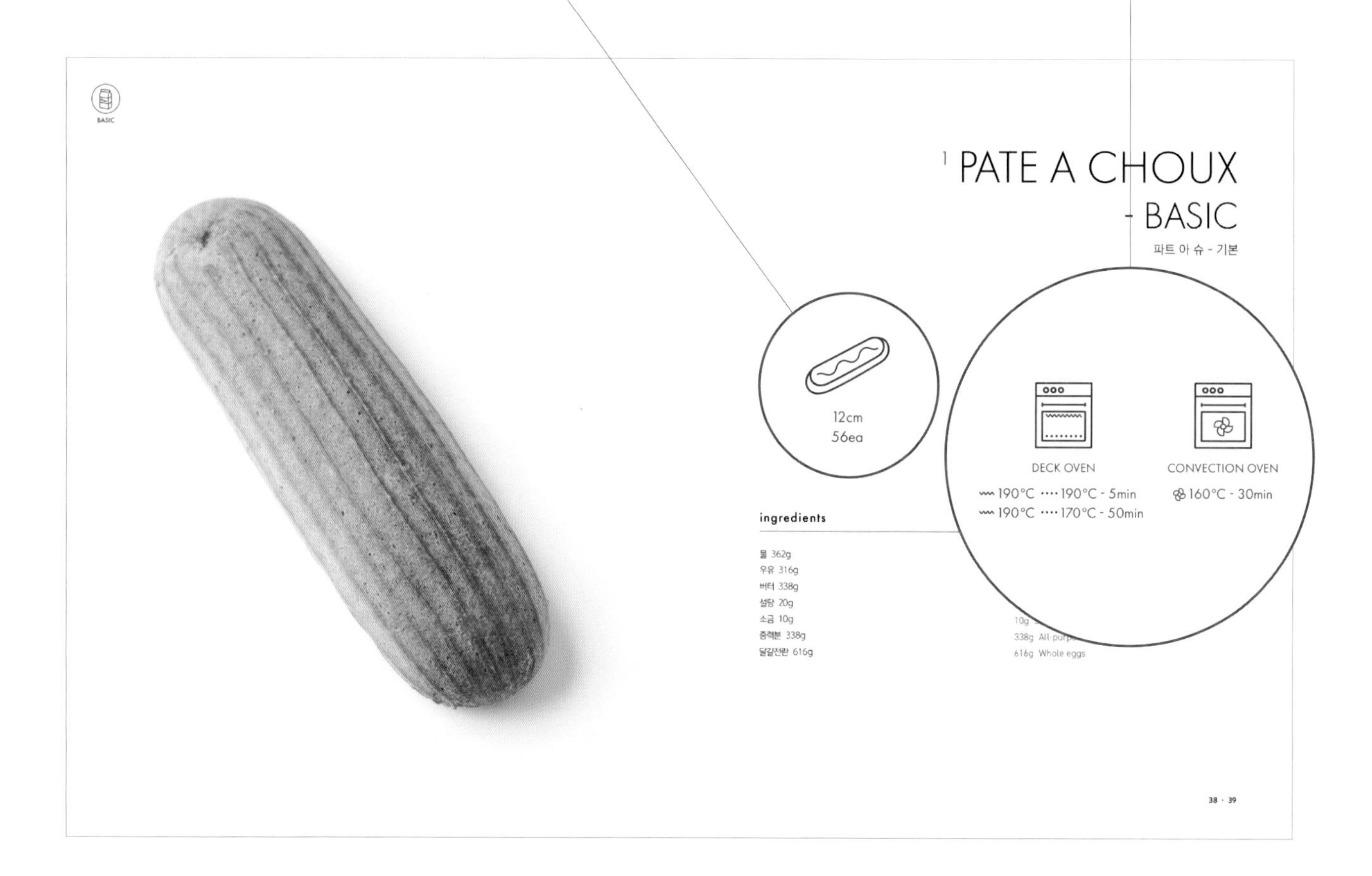

이 책에서는 밀가루를 사용하는 기본 슈 반죽, 쌀가루를 사용하는 글루텐 프리 슈 반죽 두 가지를 소개합니다. 기호에 따라 원하는 반죽을 선택할 수 있습니다. 기본 슈 반죽 레시피는 39p, 글루텐 프리 슈 반죽 레시피는 43p에서 확인하실 수 있습니다.

In this book, two kinds of choux recipes are indicated: Basic choux dough using wheat flour and gluten-free choux dough using rice flour. You can choose the dough you like according to your preference. The basic choux recipe can be found on page 39, and the gluten-free choux recipe on page 43.

12cm 에클레어 기준 15개가 완성되는 배합입니다. 필요한 만큼 배합을 줄이거나 늘려 사용할 수 있습니다.

The choux recipe yields 15 pieces of 12cm éclair. The recipe can be divided or multiplied as needed.

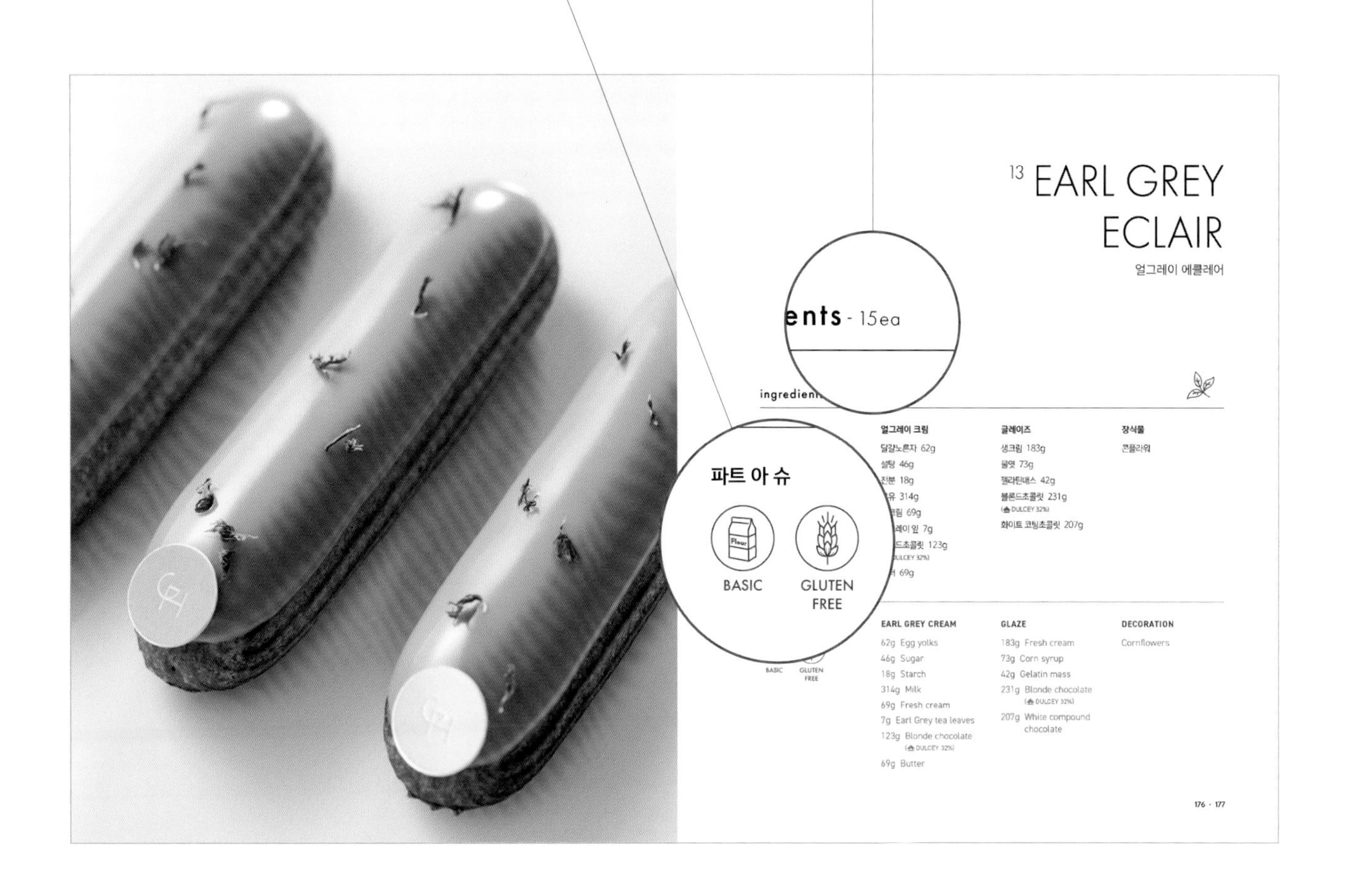

CONTENTS

Preparation

Basic

Fruits

Nut & Chocolate

Aroma

Cream & Milk

Special Recipe

Preparation

CHOUX & ECLAIR

INGREDIENTS

TOOLS

THE ROLE OF INGREDIENTS

INGREDIENTS FOR GLUTEN FREE RECIPES

UNDERSTANDING PATE A CHOUX

CONVECTION OVEN AND DECK OVEN

슈 & 에클레어 Choux & Éclair

슈(choux)는 프랑스어로 '양배추'라는 뜻을 가지고 있습니다.
슈 반죽은 다량의 수분을 함유하고 있어 오븐 안에서 많은 팽창이 일어납니다. 완성된 반죽을 오븐에 넣고 일정 시간이 지나면 오븐의 열기로 인해 슈 표면이 건조되면서 막이 형성됩니다. 시간이 지나 표면 막이 단단해지게 되면 더이상 반죽이 늘어나지 않아 표면이 갈라지기 시작하고 갈라진 틈으로 반죽이 계속해서 팽창하면서 울퉁불퉁한 모양으로 완성되는데 이 모양이 양배추를 닮았다고 해서 '슈'라는 이름이 붙여진 과자입니다.
현대에는 슈에 더해지는 장식이 화려해지면서 표면이 갈라지지 않게 반죽을 굽는 것을 선호하게 되었습니다. 반죽을 성형한 후 슈거파우더를 도포해 굽게 되면 반죽이 오븐 안에서 충분히 팽창하는 동안 슈 표면이 건조되지 않아 갈라지지 않는 매끈한 슈가 완성됩니다.

에클레어는 '번개'라는 뜻을 가지고 있습니다.
에클레어의 맛이 너무 좋아 번개처럼 빨리 먹어치운다고 하여 붙여진 이름입니다. 슈와 동일한 반죽을 사용하지만 에클레어는 길쭉하게, 슈는 동그랗게 성형해 구워낸다는 점이 다릅니다. 에클레어 외에도 슈 반죽으로 만들 수 있는 과자는 매우 다양합니다.

Choux (*singular*, chou) means 'cabbage' in French.

Choux paste has a high content of moisture, resulting in a great deal of expansion in the oven. After placing the finished dough in the oven, the surface begins to dry within a certain period of time due to the heat in the oven, forming a dry outer layer. As it continues to bake, the outer layer becomes harder, the dough no longer stretches, causing the surface to crack, and the dough continues to expand between the cracked edges, forming an uneven surface. Because the finished shape resembles a cabbage, thus the pastry was named 'choux.'

In modern days, it is preferred to bake choux without cracking the surface as the decorations added to the choux become more glamorous. When the piped dough is baked with powdered sugar dusted on top, the surface does not dry, causing the surface not to crack while the dough expands sufficiently in the oven.

Éclair means 'lighting.'

It is named so because éclair tastes so good that it's eaten away fast, like the speed of lighting. It uses the same dough as choux, but the difference is that the éclair is piped longer, and the choux are piped round. In addition to éclair, there are varieties of pastries that can be made from the choux dough.

재료 Ingredients
우유 (MILK)
생크림 (FRESH CREAM)
달걀 (EGGS)
소금 (SALT)
바닐라빈 (VANILLA BEAN)
버터 (BUTTER)

달걀

포장지에 적힌 산란일자와 포장일을 확인하여 신선한 달걀을 구입하는 것이 중요합니다. 달걀을 사용하기 전에는 물로 깨끗이 씻어 표면에 묻은 이물질을 제거한 후 사용하는 것을 권장합니다. 단, 물로 씻은 달걀을 보관하는 경우에는 세균이나 바이러스 균이 침투할 수 있으므로 필요한 만큼만 세척하여 사용하는 것이 좋습니다.

우유

우유는 제과에 있어 제품의 영양가를 높여줍니다. 또한 단백질과 유당을 함유해 메일라드 반응을 촉진시켜 과자 껍질의 색을 좋게 만듭니다. 고지방 우유, 저지방 우유, 칼슘 강화 우유처럼 특정 영양소를 강화한 우유, 유당을 소화하지 못하는 사람을 위한 락토프리 우유 등 다양한 제품이 있습니다. 이 책에서 사용한 우유는 모두 일반 우유입니다.

생크림

생크림은 식물성 기름이 첨가되지 않은 신선한 것을 구입해 유효기간 내에 사용하는 것이 좋습니다. 식물성 기름을 혼합한 가공 생크림은 작업성을 높여주고 보관 기간이 길지만 입안에서 잘 녹지 않고 풍미가 떨어지기 때문에 권장하지 않습니다. 생크림은 유지방 함량 20% 정도의 저지방 생크림부터 유지방 함량 45% 전후의 고지방 생크림까지 매우 다양합니다. 이책에서는 유지방 함량 38%의 동물성 생크림을 사용하였습니다.

소금

이 책에서 사용한 소금은 프랑스 서부, 대서양을 마주보는 게랑드 일대에서 생산되는 소금입니다. 염전 위에 꽃처럼 피어난 소금 결정을 하나하나 손으로 수확하여 '소금의 꽃(플뢰르 드 셀 fleur de sel)'이라 불립니다. 짠맛 뿐만 아니라 감칠맛까지 느낄 수 있으며 캐러멜과 같은 단맛과도 잘 어우러집니다.

바닐라빈

바닐라빈은 우유, 생크림, 버터와 같은 유제품 향의 조화를 높여주며 달걀 특유의 비릿함을 잡아줍니다. 긁어낸 바닐라빈은 크림이나 반죽에 넣어 향을 내고, 사용하고 남은 껍질은 설탕에 넣어 두었다가 설탕과 함께 푸드프로세서에 갈아 바닐라슈거를 만들어 사용할 수 있습니다.

버터

제과의 기본 재료인 버터는 냉동고나 냉장고에 넣어 차가운 상태, 실온에 꺼내두어 말랑한 상태, 전자레인지에 녹여 완전히 풀어진 상태 세 가지 형태로 사용합니다. 버터는 유지방 함량이 80% 이상인 퓨어버터를 사용하는 것이 좋습니다. 산화되기 쉬운 퓨어버터는 신선한 것을 구입해 빠른 시일 안에 사용하고, 오래 두고 사용해야 할 경우에는 냄새를 흡수하지 않도록 잘 밀폐하여 냉동 보관합니다. 이 책에서 사용한 버터는 제조 공정 중 유산균을 첨가해 숙성한 프랑스 천연 발효 버터입니다. 느끼함이 없고 유지방이 응축된 고소하고 담백한 맛을 느낄 수 있습니다.

Eggs

It is important to purchase fresh eggs by checking the date eggs were laid* and the packaging date printed on the carton. It is recommended to wash them thoroughly with water before using them to remove foreign substances on the surface. However, there is a possibility that bacteria or viruses may penetrate while storing eggs washed with water, so it's better to wash them only as needed.

* May not be indicated in some countries.

Milk

Milk improves the nutritional value of the baked goods in confectionery. It also contains protein and lactose, which promotes the Maillard reaction that helps to make an appetizing outer color of the confectionery goods. There are various kinds of milk, which include high-fat milk, low-fat milk, and milk with certain nutrients strengthened such as calcium-fortified milk, and lactofree milk for those who cannot digest lactose. The milk used in this book is regular whole milk.

Fresh cream (whipping cream)

It is recommended to purchase fresh cream that does not contain vegetable oils and use it within the expiration date. Processed whipping cream blended with vegetable oils increases workability and has a long shelf life; however, it does not melt nicely in the mouth and has poor flavor; therefore, it's not recommended. There are varieties of fresh creams ranging from low-fat fresh cream with a milk fat content of about 20% to high-fat fresh cream with a milk fat content of around 45%. In this book, we used fresh dairy cream containing 38% of milk fat.

Salt

The salt used in this book is produced in the region of Guerande, facing the Atlantic Ocean in western France. Each salt crystal blooming like flowers on the salt field are harvested by hand is called 'flower of salt (fleur de del).' You can feel not only its saltiness but also savory taste, and pairs well with sweet flavors such as caramels.

Vanilla beans

Vanilla bean enhances the harmony of dairy flavors such as milk, fresh cream, and butter and helps reduce the peculiar taste of eggs some may find fishy. The scraped vanilla bean seeds can be added to creams or batters to enhance flavor, and the remaining skin can be stored inside sugar, which can be ground all together with a food processor to make vanilla sugar.

Butter

Butter, which is the basis of confectionery ingredients, is used in three forms: cold state by keeping in a freezer or refrigerator, soft state by keeping in ambient temperature, and completely dissolved state by melting in a microwave. As for the butter, pure butter containing more than 80% of milk fat would be recommended to use. Pure butter should be purchased fresh and used as soon as possible because it's easy to oxidize. If it needs to be stored for a long time, it should be sealed tight, not to absorb any odor, and kept frozen. The butter used for this book is naturally fermented butter from France, aged by adding lactic acid bacteria during the manufacturing process. It doesn't taste greasy and gives a savory and clean flavor from concentrated milk fat.

럼 (RUM)
마스카르포네 치즈
(MASCARPONE CHEESE)
마롱 페이스트
(MARRON PASTE)
젤라틴 (GELATIN)
펙틴 (PECTIN)
젤라틴매스 (GELATIN MASS)
아가아가 (AGAR-AGAR)

마스카르포네 치즈

이탈리아의 대표적인 디저트 '티라미수'를 만드는 데 있어 필수적인 재료입니다. 이탈리아의 대표 치즈이며 지방 함량이 높아 섬세하고 부드러우면서도 진한 크림의 향을 느낄 수 있습니다.

럼

당밀과 사탕수수를 원료로 만든 증류주로 크림이나 과자의 풍미를 돋우는 데 도움을 줍니다.

마롱 페이스트

페이스트 상태로 만든 밤에 당을 첨가해 만듭니다. 단맛과 고소함을 동시에 느낄 수 있는 재료로, 당을 넣지 않은 마롱 퓌레부터 당도가 높은 마롱 크림까지 다양한 제품이 있으며, 제조사에 따라 당도의 차이가 있으므로 기호와 용도에 따라 선택할 수 있습니다.

펙틴

펙틴은 사과 찌꺼기나 감귤류의 껍질에서 추출한 천연 다당류입니다. 주로 크림이나 젤리의 질감을 완성하거나 안정화하는 데 사용되며 제조 과정 중에 음식물의 형체(바디감)와 질감(텍스처)을 만들어 주는 역할을 합니다. 펙틴은 일반적으로 하이 메톡실(HM), 로우 메톡실(LM) 두 가지 유형으로 나눌수 있으며 시중에는 다양한 종류의 펙틴이 존재합니다. 각각의 펙틴마다 물리화학적 특성이 다르므로 작업 전 공급자에게 제품 정보를 확인하는 것이 좋습니다.

젤라틴 & 젤라틴매스

응고제의 일종인 젤라틴은 무스나 젤리 등을 굳히는 데 사용되는 재료입니다. 소나 돼지의 뼈 또는 껍질에서 추출한 콜라겐을 정제해 건조시켜 만듭니다. 가루 타입과 얇은 판 형태의 제품이 있습니다. 이 책에서는 가루 형태의 젤라틴(200bloom)을 5배의 물과 혼합해 굳힌 후 사용하였습니다.(53p)

아가아가

한천을 주 원료로 하는 겔화제로 단단하면서도 탄력이 적은 겔을 형성할 수 있으며 겔화된 상태에서 다시 가열해 사용하는 것이 가능합니다.

Mascarpone cheese

This is an essential ingredient in making the classic Italian dessert called 'Tiramisu.' It is the representative cheese of Italy, which has a high-fat content, giving a delicate and soft yet rich creamy aroma.

Rum

Rum is a distilled liquor made from molasses and sugar cane, which helps enhance the flavor of cream or sweets.

Marron paste

It is made by adding sugar to mashed chestnuts. As an ingredient that has both sweet and nutty flavors at the same time, there are various products available, from marron puree without sugar to marron cream with high sugar content. There are differences in sugar content depending on the manufacturer, which can be chosen to use according to one's preference.

Pectin

Pectin is a natural polysaccharide extracted from apple pomace or citrus peels. It is usaually used to complete or stabilize the texture of cream or jelly and at the same time to complete the shape (body structure) and texture (sensory) of the product. Generally, there are two types of pectin; high methoxyl (HM) and low methoxyl (LM), and there are more types available in the market. Since each pectin has different physicochemical properties, it is recommended to check with the supplier for information about the product before use.

Gelatin & gelatin mass

Gelatine is a kind of coagulant that is used to harden mousse or jelly. It is made by refining and drying collagen extracted from bones or skins of cattle or pigs. There are powder type and thin sheet type of gelatin. For this book, the powder type of gelatin (200 bloom) was mixed with five times of water and hardened to use (gelatin mass). (53p)

Agar-agar

A gelling agent, typically using algae as the main ingredient, is used to form a firm and less elastic gel and can be reheated to use again.

도구 Tools

푸드프로세서
(FOOD PROCESSOR)

짤주머니 (PIPING BAG)

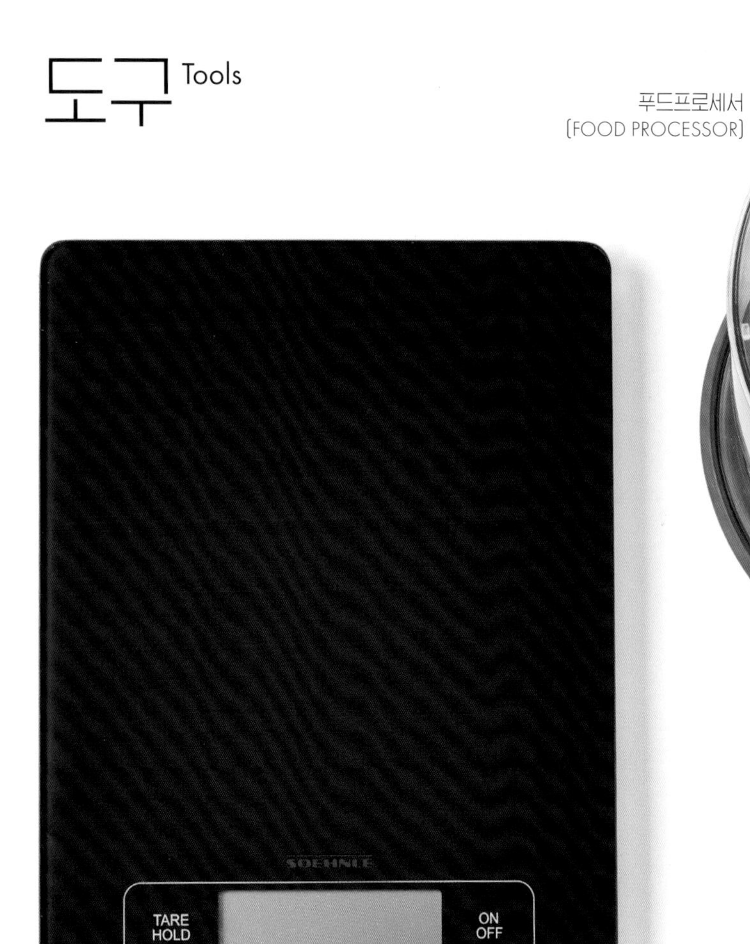

저울 (SCALE)

깍지 (TIPS)

핸드블렌더
(HAND BLENDER)

스크래퍼
(SCRAPER)

믹싱볼 (MIXING BOWLS)

스패출러 (SPATULA)

저울

베이킹에 있어 정확한 계량은 매우 중요합니다. 1g 차이로도 맛이 달라지거나 제품이 아예 만들어지지 않을 수도 있기 때문입니다. 따라서 눈금저울보다는 1g 단위로 표시되는 전자저울을 사용하는 것이 좋습니다. 재료를 담는 볼을 올리고 0을 맞춰 0점 조절을 한 다음 계량을 시작합니다. 1g보다 더 적은 양을 계량한다면 1g에서 1/2, 1/4과 같이 줄여나가면 됩니다.

푸드프로세서

재료를 잘게 자르고 다지는 용도로 사용합니다.

짤주머니/깍지

반죽이나 크림을 담아 짜거나 모양을 내기 위해 사용하는 도구입니다. 굽는 반죽의 경우 천 재질의 짤주머니를 사용해도 좋지만, 가열하는 과정 없이 바로 먹는 크림 종류의 경우에는 식중독 예방을 위해 일회용 비닐 짤주머니를 사용하고 재사용하지 않는 것이 위생적입니다. 다양한 종류의 깍지를 짤주머니에 끼워 용도에 맞게 사용합니다.

믹싱볼

반죽을 섞거나 거품을 올리는 작업에 사용되는 필수 도구입니다. 스테인리스, 유리, 폴리카보네이트 등 다양한 재질이 있으며 용도에 맞춰 큰 사이즈(지름 25cm)부터 작은 사이즈(17cm)까지 3~4개 정도 여유롭게 준비하는 것이 좋습니다. 스테인리스 재질은 가볍고 열 전달이 빠른 반면, 유리와 폴리카보네이트 재질은 상대적으로 열 전달이 느려 뜨거운 것이나 차가운 것을 보다 오래 유지시켜줍니다.

핸드블렌더

글레이즈, 크림, 가나슈 등을 균일하게 혼합하는 데 유용한 도구입니다.

스패출러

반죽이나 크림을 펼치거나 고르게 정리할 때 사용합니다. 손잡이와 날이 일직선인 것, L자로 굽은 것 두 가지 종류가 있습니다.

스크래퍼

넓은 면적을 이용해 반죽을 이기거나 혼합하는 데 사용하며, 반죽을 떠 담기에도 유용한 도구입니다.

Scale

Accurate scaling is extremely important in baking. It's because the taste may change, or the product may not be made at all, even with a 1 gram of difference. Therefore, it is better to use an electronic scale that displays in 1-gram increments rather than a needle-type scale. Place a bowl for scaling the ingredient on the scale and set to 0 (tare), and start weighing. If you measure less than 1 gram, reduce from 1 gram to 1/2, 1/4, and so on.

Food processor

This is used to chop and grind ingredients.

Piping bags/tips (nozzles)

These are used for piping or shaping batter or cream. Cloth material piping bags may be used for batters that will be baked. But for creams consumed without the cooking process, it is recommended to use disposable plastic piping bags for hygienic reasons and not to re-use them to prevent food poisoning. Use various types of tips with piping bags accordingly.

Mixing bowls

It is an indispensable tool used for mixing doughs or whipping. There are various materials such as stainless steel, glass, polycarbonate, etc. It is recommended to prepare 3~4 bowls, from large size (25cm diameter) to small size (17cm diameter), to use according to the purpose. Stainless steel material is light and transfers heat faster, while glass and polycarbonate materials have relatively slower heat transfer, which helps keep hots or colds longer.

Hand blender

A hand blender is a useful tool for mixing glazes, creams, and ganache evenly.

Spatula

It is used to spread or evenly organize cream or batter. There are two types: one that's straight from the handle to the blade and one in which the blade is bent to L-shape.

Scraper

A scraper is used to beat or mix batters and is also a useful tool for transferring batter.

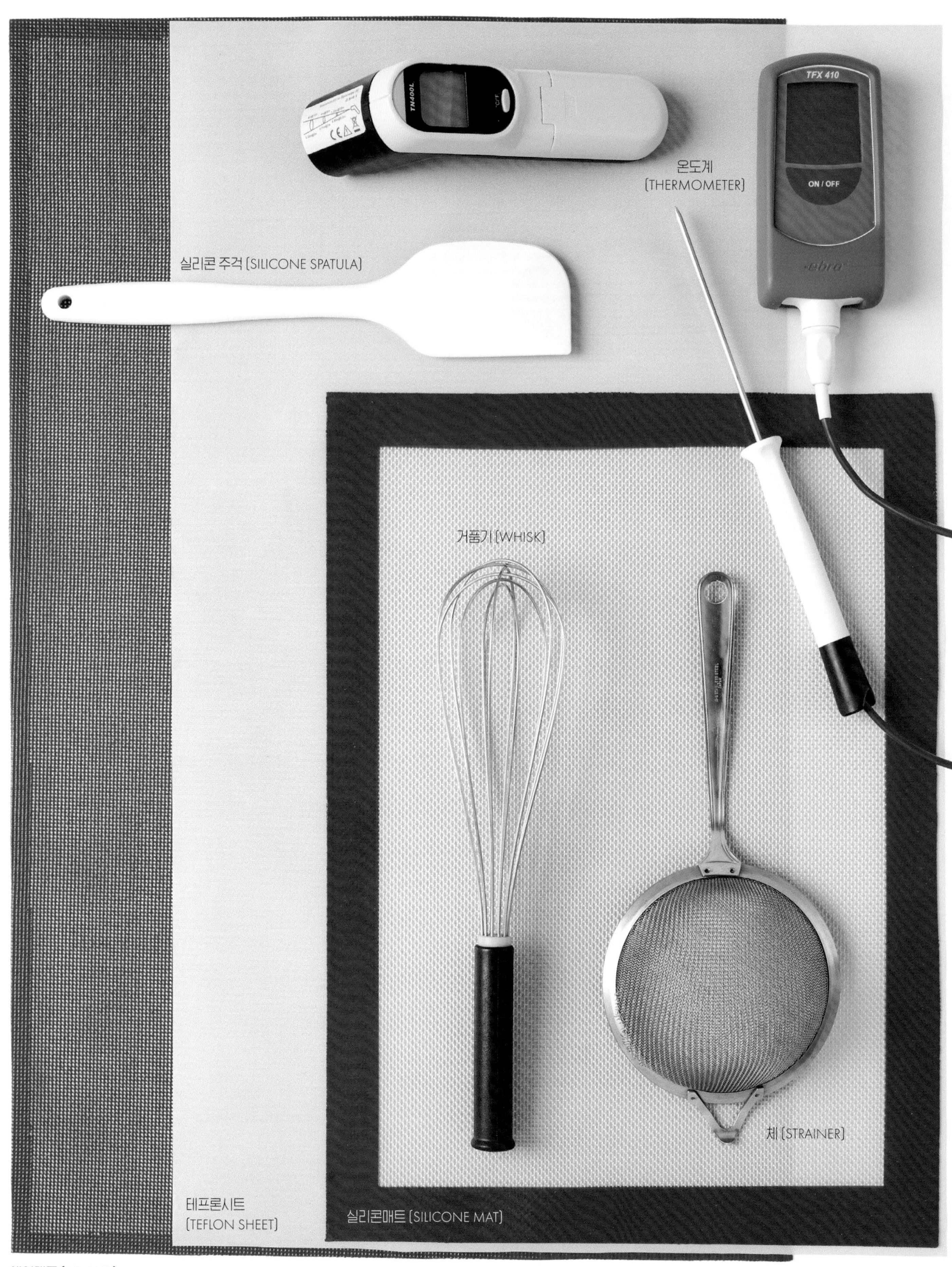
TN400L
TFX 410
ON / OFF
ebro
온도계
〔THERMOMETER〕
실리콘 주걱 〔SILICONE SPATULA〕
거품기 〔WHISK〕
체 〔STRAINER〕
테프론시트
〔TEFLON SHEET〕
실리콘매트 〔SILICONE MAT〕
에어매트 〔AIR MAT〕

오븐용 페이퍼

오븐용 페이퍼는 유산지, 테프론시트, 실리콘매트 등의 제품을 말하며 철판에 깔아 제품과 철판이 달라붙는 것을 방지하고 구웠을 때 제품의 밑면이 깔끔하게 분리되게 하는 역할을 합니다. 실리콘 페이퍼, 내열 페이퍼, 방수 페이퍼 등의 특수 가공 처리를 한 제품들도 있습니다.

에어매트

그물 모양으로 만들어진 오븐용 실리콘매트입니다. 촘촘하게 구멍이 뚫려 있기 때문에 오븐 안에서 공기의 흐름이 좋고 반죽이 미끄러지지 않아 모양이 그대로 유지됩니다.

온도계

정확한 계량 못지않게 베이킹에서 중요한 것이 바로 온도입니다. 온도계는 주로 반죽이나 크림의 온도를 체크하는 데 사용되며 고온의 시럽을 만들 때에도 사용할 수 있도록 250℃까지 측정이 되는 제품을 선택하는 것이 좋습니다. 표면의 온도를 측정하는 비접촉식 적외선 온도계와 내용물 안에 넣어 온도를 측정하는 접촉식 온도계 두 가지 종류가 있습니다.

실리콘 주걱

볼에 담긴 반죽이나 크림을 섞거나 모으는 데 사용합니다. 힘 없이 휘어지는 것보다는 어느 정도 탄력이 있는 제품이 사용하기 더 편리합니다. 주걱 앞부분부터 손잡이 끝부분까지 홈이 없이 매끄럽게 연결되어 만들어진 주걱이 이물질이 끼지 않아 위생적으로 더 좋습니다.

거품기

재료를 섞어 혼합하거나 공기를 포집할 때 사용하는 도구입니다. 와이어 부분이 유연하면서도 튼튼하며 손으로 잡았을 때 편한 것을 고르는 것이 좋습니다.

체

가루 재료를 곱게 내리거나 섞어 놓은 액체 재료를 거르는 데 사용하는 도구로 다양한 사이즈가 있습니다. 제과에서 가루 재료들은 모두 체에 내린 후 사용하는 것이 기본입니다. 가루 재료를 체에 내리면 뭉쳐 있던 멍울은 풀어지고, 불순물이 제거되며, 체에 내리는 과정에서 가루와 가루 사이에 공기가 혼입되어 다른 재료와 더 잘 섞이게 됩니다.

Papers for oven

Papers for the oven refers to parchment paper, Teflon sheets, and silicone mats. These are to be lined on a baking tray to prevent products from sticking and to help the bottom of the baked goods separate neatly. There are also specially processed products such as silicone paper, heat-resistant paper, and waterproof paper.

Air mat

This is a net-shaped silicone mat for ovens. Because of its fine holes, the air in the oven flows through and around, and the dough does not slip on it, which helps to maintain its shape.

Thermometer

Temperature is every bit as important as accurate scaling when it comes to baking. Thermometers are mainly used to check the temperature of dough/batter or cream and check the temperature of high-temperature syrups. Therefore, it is recommended to select a product that can measure up to 250°C. There are two types: a non-contact type infrared thermometer that measures surface temperature and a contact type thermometer that measures the temperature by placing it inside the content.

Silicone Spatulas

These are used to mix or collect batter or cream in the bowl. Spatula with a resilient blade is more convenient to use than the blades that bend easily. The spatula that is smooth all the way, and has no groove from the blade to the tip of the handle, is better in the means of hygiene because no foreign substance will remain in between.

Whisk

This is a tool used to mix ingredients or to collect air. Choosing a whisk with wires that are flexible yet sturdy and comfortable when held by hand is good to use.

Strainer

A strainer is used to sift powder ingredients a fine form or filter mixture of liquid ingredients, and there are various sizes available. In confectionery, it is a ground rule to sieve all powdered ingredients. When the powdered ingredients are sifted, the lumps break up, impurities are removed, and the air gets to mix in between the particles, which helps to better mix with other ingredients.

재료의 역할

The role of Ingredients

유지(버터)

- 유지는 필요 이상의 글루텐 형성을 억제해 반죽이 잘 늘어나도록 하고 반죽에 유연성을 줍니다.
  * 글루텐이 과잉 생성되면 반죽의 점탄성이 너무 강해져 반죽이 부풀지 않고 단단해집니다.

수분

- 반죽에 함유된 수분은 오븐에서 열에 의해 증기로 바뀌면서 반죽을 팽창시키는 역할을 합니다.
- 밀가루 전분의 호화와 글루텐 형성에 필요한 필수 요소입니다.

밀가루

- 밀가루의 전분과 수분이 만나 가열의 과정을 거치면 호화된 전분 입자가 됩니다. 전분이 호화되면 반죽이 잘 늘어나게 됩니다.
- 밀가루의 단백질과 수분이 만나 반죽의 과정을 거치면 글루텐이 생성됩니다. 글루텐은 반죽에 점성을 주고 부푼 모양을 유지하도록 해줍니다.

달걀

- 달걀 노른자에 레시틴이 서로 잘 섞이지 않는 유지와 수분을 잘 섞이게 하는 유화작용을 합니다.
- 반죽의 되기를 조절합니다.
- 오븐의 열에 의해 달걀 단백질이 응고되면서 부푼 모양을 유지하는 데 도움을 줍니다.

Milk fat (Butter)

- Milk fat suppresses the excess formation of gluten to help the dough stretch better and gives the dough flexibility.
  * When excess gluten is formed, the viscoelasticity of the dough becomes too strong; the dough will not rise enough and become hard.

Water

- Water contained in the dough is converted into steam by heat in the oven, which helps the dough expand.
- It is an essential element for the gelatinization of starch in the flour and development of gluten.

Flour

- When the water and starch in the flour are mixed and go through the process of heating, it turns to gelatinized starch particles. When the starch is gelatinized, the dough stretches better.
- Gluten is produced when the protein in the flour meets water and goes through the mixing process. Gluten gives the dough a viscous consistency and helps to keep its expanded shape.

Eggs

- The lecithin in egg yolks helps emulsify the fat and water that do not mix well together.
- Helps control the consistency of the dough.
- It helps to maintain the expanded shape as the egg protein solidifies by the heat of the oven.

글루텐 프리 레시피를 위한 재료들

Ingredients for Gluten Free Recipes

디저트 샵을 운영하다보면 알레르기로 인해 글루텐 섭취가 어려운 분들을 종종 만나게 됩니다. 이런 분들을 보며 팀 가루하루의 호기심과 고민이 시작되었고, 수많은 테스트를 거쳐 지금의 글루텐 프리 슈 레시피를 완성할 수 있었습니다.
이 책에서 소개하는 글루텐 프리 슈 레시피를 활용하면 글루텐 알레르기가 있으신 분들은 물론 건강상의 이유로 글루텐 섭취를 제한하고 싶은 분들까지도 맛있고 건강한 가루하루의 에클레어를 즐기실 수 있습니다.

1. 쌀가루

시판되는 쌀가루의 종류

쌀가루는 침지(불림) 과정의 유무에 따라 '건식쌀가루'와 '습식쌀가루'로 구분되며 각각의 제조 과정은 아래와 같습니다.

- 건식쌀가루의 제조 과정

세척	탈수	제분	건조

- 습식쌀가루의 제조 과정

세척	침지(불림)	탈수	제분	건조

침지 과정은 쌀 원물을 물에 불려 전분 입자 사이의 간격을 벌어지게 해 스펀지와 같은 조직을 만드는 과정입니다. 상온에서 유통되는 습식 쌀가루의 경우 이렇게 쌀 전분 입자를 벌려 놓은 상태에서 건조시키므로(상온 유통 과정 중 미생물이 번식하는 것을 방지하기 위해) 순간적으로 많은 양의 수분을 흡수할 수 있게 됩니다.
제과에서 밀가루 대체 재료로 쌀가루를 사용할 경우에는 수분 흡수와 보유력이 좋은 습식쌀가루를 사용하는 것이 적합합니다. 건식쌀가루의 경우 반죽 과정에서 쌀 전분이 충분한 수분을 흡수할 수 없어 반죽이 질어지고 퍼지는 요인이 되기 때문입니다. 또한 입자가 고울수록 부드러운 식감을 주기 때문에 기호나 용도에 따라 쌀가루의 입자 크기를 선택해 사용할 수 있습니다. 시판되고 있는 박력쌀가루의 경우 부드러운 식감을 위해 밀가루와 비슷한 입자로 제분한 것입니다.

When you run a dessert shop, you will often meet people who have difficulty ingesting gluten due to allergies. As I watched these people, Team GARUHARU's curiosity and concerns began, and after numerous tests, we were able to complete the current gluten-free choux recipe.
By using the gluten-free recipe introduced in this book, you can enjoy GARUHARU's delicious and healthy éclair, not only for those with gluten allergies but also for those who want to limit gluten intake for health reasons. We hope you will enjoy the various styles of éclairs by choosing recipes presented in this book.

1. Rice flour

Types of rice flour available in the market

Rice flour is divided into 'dry(-milled) rice flour' and 'wet(-milled) rice flour' depending on whether or not it went through the process of immersion (soaking), and each manufacturing process is as follows.

• Manufacturing process of dry rice flour

• Manufacturing process of wet rice flour

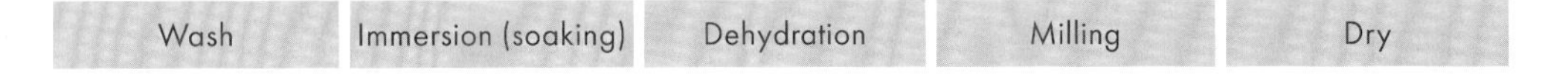

The immersion is a process in which raw rice grains are soaked in water to open the gaps between starch particles to create a sponge-like tissue. In the case of the wet rice flours distributed at ambient temperature, rice starch particles are dried in an open state (to prevent microorganisms from multiplying during the distribution process in ambient temperature) so that a large amount of moisture can be absorbed instantaneously.
When using rice flour to substitute wheat flour in confectionery, it is suitable to use wet rice flour, which has good moisture absorption and retention. This is because, when dry rice flour is used, the rice starch cannot absorb enough moisture during the kneading process, which can cause the dough to become thin and run easily. Additionally, the finer the particles, the softer the texture, so you can select the particle size of rice flour and use it according to your taste or purpose. The commercially available soft rice flour is ground into particles similar to wheat flour for a smooth texture.

❶ ❷ ❸

쌀가루 만들기

❶ 쌀을 흐르는 물에 3~4회 깨끗이 세척한다.
❷ 냉장고에서 12시간 정도 불린다.
❸ 체에 받쳐 30분간 탈수한다.
❹ 45℃에서 10시간 정도 건조시킨다.
❺ 푸드프로세서에 넣고 고운 가루 상태가 될 때까지 갈아준다.

2. 잔탄검

잔탄검은 글루텐 프리 슈 반죽이 가지지 못하는 탄성과 점도를 더해줍니다. 이러한 잔탄 검의 특성은 글루텐 프리 슈 반죽이 잘 늘어나게 도와주며 굽는 과정 중에 생성되는 증기를 잡아주어 반죽이 더 잘 부풀게 해줍니다. 잔탄검이 없는 경우 생략해도 무방하지만 레시피에 제시된 분량 만큼 첨가할 경우 더욱 볼륨감 있는 글루텐프리 슈를 완성할 수 있습니다.

❹ ❺-1 ❺-2

How to make rice flour

❶ Wash rice 3~4 times under running water.

❷ Let it soak in the refrigerator for about 12 hours.

❸ Strain and dehydrate for 30 minutes.

❹ Dry at 45°C for 10 hours.

❺ Grind in a food processor until it turns into a fine powder.

2. Xanthan gum

Xanthan gum adds elasticity and viscosity that gluten-free choux paste does not have. This property of xanthan gum helps the gluten-free choux paste stretch better and inflate more by trapping the vapors produced during the baking process. If you don't have xanthan gum, you can omit it, but if you add the suggested quantity in the recipe, you can make more voluminous gluten-free choux.

파트 아 슈의 이해

Understanding PATE A CHOUX

슈는 내부에 빈 공간이 있는 독특한 구조의 과자입니다.
이러한 구조는 전분의 호화와 수증기압을 통해 만들어지게 됩니다. 반죽 과정 중 밀가루의 전분은 수분과 만나 가열 과정을 거쳐 호화되며, 호화된 전분은 점도가 생겨 잘 늘어나게 됩니다. 굽는 과정 중 반죽 속 수분은 오븐 열에 의해 가열되어 증기로 바뀌면서 반죽을 부풀려 팽창시킵니다.

수분 Water
호화된 반죽 Gelatinized dough
수증기 Steam

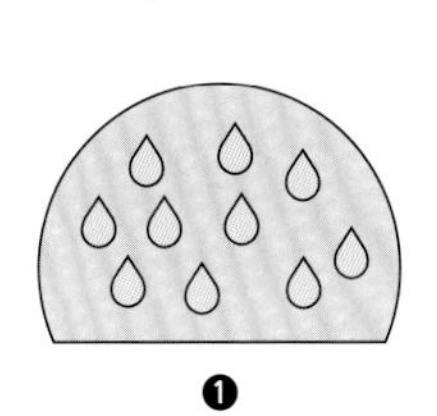
❶

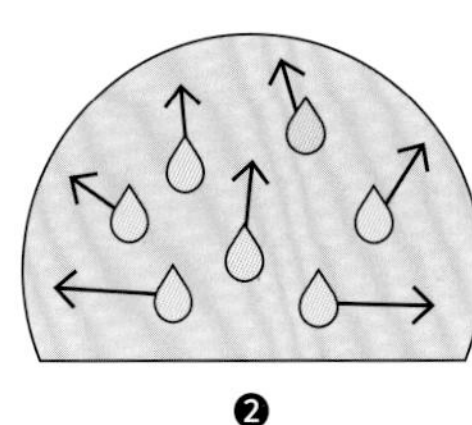
❷

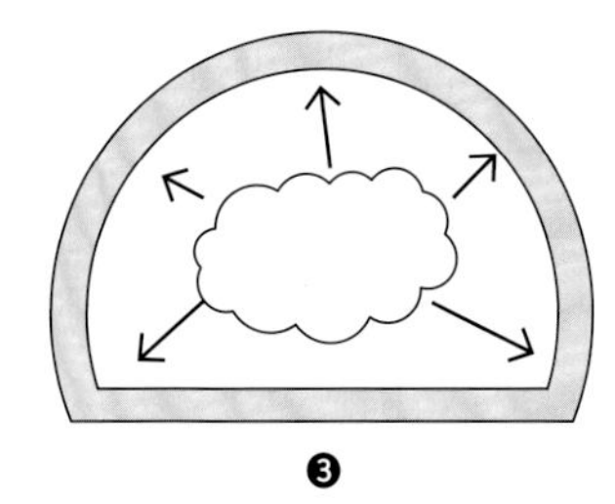
❸

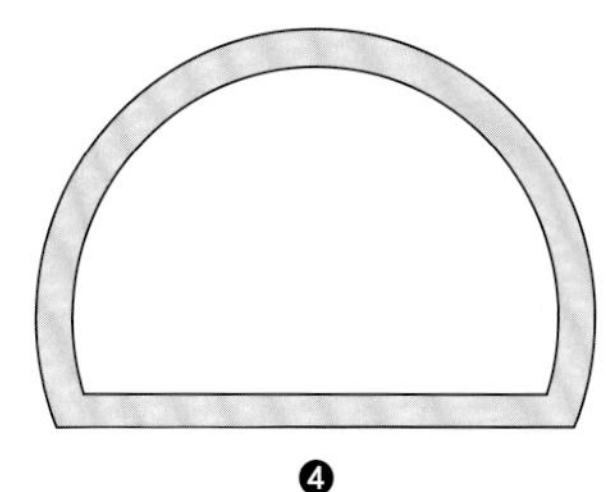
❹

❶ 오븐의 열이 반죽 외부에서 중심을 향해 전달되어 반죽 내부의 수분이 따뜻해지기 시작합니다.
❷ 수분의 온도가 점점 상승하고 반죽 중심부에서 수증기가 발생하기 시작하면서 반죽을 부풀려 팽창시킵니다.
❸ 반죽 표면이 건조되기 시작해 표면 막이 형성되고, 이 막이 증기를 내부에 가두면서 내부 공간이 형성됩니다.
❹ 반죽이 건조되어 단단해지며, 달걀 단백질 등의 열응고에 의해 부푼 형태가 그대로 유지됩니다.

Choux is a unique pastry with a hollow space inside.
This structure is created by starch gelatinization and water vapor pressure. During the kneading process, the starch in the flour gets mixed with water and then undergoes a heating process, then the gelatinized starch becomes viscous, helping to stretch well. During the baking process, the moisture in the dough is heated by the heat in the oven and turns into steam, which helps to inflate and expand the dough.

❶ The heat from the oven gets transferred from the outside of the dough towards the center, and the water in the dough starts to get warm.
❷ As the temperature of the water gradually increases, vapor begins to develop in the center of the dough, and the dough starts to inflate and expand.
❸ As the surface of the dough starts to dry, a layer is formed on the surface, and as this layer traps steam inside, a hollow interior is formed.
❹ The dough continues to get dried and become hardened, which makes the puffed form retain by heat coagulation of egg protein, etc.

컨벡션 오븐과 데크 오븐

Convection Oven and Deck Oven

오븐의 종류는 '데크 오븐'과 '컨벡션 오븐'으로 나눌 수 있습니다.
데크 오븐은 오븐의 상부와 하부에 위치한 열선을 통해 열을 전달하며, 컨벡션 오븐은 오븐 내부의 팬이 오븐 안의 열을 순환시키는 방식으로 열을 전달합니다.
컨벡션 오븐의 경우 뜨거운 바람으로 열을 전달하기 때문에 다단 조리가 가능한 장점이 있지만 반죽이 건조해지기 쉽습니다. 따라서 컨벡션 오븐은 바삭한 식감의 쿠키나 파이 등을 굽기에 적합하며, 촉촉한 식감의 제품을 구울 때는 적합하지 않습니다.
슈의 경우 데크 오븐과 컨벡션 오븐을 모두 사용할 수 있습니다. 다만 컨벡션 오븐에 슈를 구울 경우 슈 표면이 더욱 빠르게 건조되어 균열이 생길 수 있으므로 표면이 갈라지지 않는 슈를 굽고 싶은 경우에는 데크 오븐을 선택하는 것이 좋습니다.

TIP. 데크 오븐이 없는 경우

데크 오븐이 없는 경우 대리석 판을 이용하면 컨벡션 오븐으로도 표면이 갈라지지 않는 슈를 완성할 수 있습니다. 먼저 컨벡션 오븐 안에 대리석 판을 넣고 대리석 판이 충분히 달궈질 때까지 예열합니다. 달궈진 대리석 판 위에 슈 반죽을 파이핑한 팬을 올린 후 오븐 전원을 끄고 슈 반죽이 충분히 팽창할 때까지 기다렸다가 다시 전원을 켜 완전히 익을 때까지 구워줍니다. 이는 달궈진 대리석의 열을 이용해 슈 반죽을 팽창할 수 있게 함과 동시에 오븐의 전원을 차단해 뜨거운 바람에 의해 슈 표면이 급속히 건조되는 것을 막는 원리입니다.

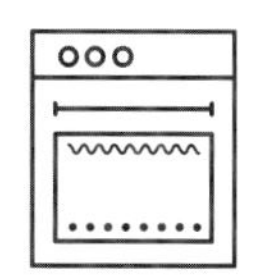

데크 오븐 DECK OVEN

상부와 하부의 열선을 통해 열을 전달

Heat distributed through the upper and lower heating wires

컨벡션 오븐 CONVECTION OVEN

팬의 바람을 통해 열을 순환

Distributes heat by circulating heated air with a fan

The types of ovens can be divided into 'deck oven' and 'convection oven.'
The deck oven distributes heat through the heating wires located at the top and bottom of the oven. As for the convection oven, heated air is distributed by the fan inside the oven circulating the heat.
Convection ovens have the advantage of being able to cook multiple trays because it transfers heat by hot air, but the dough may easily dry out. Therefore, a convection oven is ideal for baking crispy textured cookies or pies and not suitable for baking products that need to be moist.
As for choux, both deck oven and convection oven can be used. However, when using a convection oven, the surface of the choux dries out faster and may crack, so for baking choux that does not crack on the surface, the deck oven is recommended.

TIP. If you don't have deck oven

Even if you don't have a deck oven, you can still bake non-cracking choux by using a marble plate. First, preheat a marble plate in the convection oven until the marble plate is heated thoroughly. Place a tray with piped choux on the heated marble plate, turn off the oven, and wait for the choux to expand sufficiently, then turn it back on to bake until the choux are fully cooked. This is a principle that allows the choux to expand by the heated marble while at the same time cutting the power of the oven to prevent the surface from drying out rapidly by hot air.

Basic

PATE A CHOUX - BASIC

PATE A CHOUX - GLUTEN FREE

DOUGH FORMATION & BAKING

FILLING

GLAZING

GELATIN MASS

CHOCOLATE TEMPERING

CORNET

BASIC

1 PATE A CHOUX - BASIC

파트 아 슈 - 기본

12cm
56ea

DECK OVEN
190°C ···· 190°C - 5min
190°C ···· 170°C - 50min

CONVECTION OVEN
160°C - 30min

ingredients

물 362g
우유 316g
버터 338g
설탕 20g
소금 10g
중력분 338g
달걀전란 616g

362g Water
316g Milk
338g Butter
20g Sugar
10g Salt
338g All-purpose (medium) flour
616g Whole eggs

1
2-1
2-2
3
4
5
6-1
6-2

Process

1. 냄비에 물, 우유, 버터, 설탕, 소금을 넣은 다음, 버터가 완전히 녹고 액체가 끓을 때까지 가열한다.
2. 불에서 내려 체 친 중력분을 넣고 한 덩어리가 될 때까지 충분히 섞어준다.
3. 다시 불 위에 올려 반죽이 전체적으로 윤기가 나고 냄비 바닥에 얇은 막이 생길 때까지 계속해서 재빠르게 섞어주며 골고루 열을 가한다.
4. 반죽을 믹싱볼에 옮긴 후 저속으로 믹싱해 약 55~60℃로 식힌다. 달걀전란은 완전히 풀어 준비한다.
5. 준비한 달걀전란을 조금씩 나누어 가며 혼합한다.
6. 완성된 반죽은 기본 에클레어 모양을 기준으로 데크 오븐의 경우 윗불 190℃/아랫불 190℃로 5분간 굽다가 아랫불을 170℃로 낮춰 50분간 굽는다. 컨벡션 오븐의 경우 160℃로 예열한 오븐에서 30분간 굽는다.

1. In a saucepan, heat water, milk, butter, sugar, and salt, until butter is completely melted and the mixture is boiling.
2. Remove from heat, and mix in sifted all-purpose flour until it forms one dough.
3. Put the saucepan back on the heat, and keep stirring quickly to apply heat evenly until the dough is shiny and forms a thin film on the bottom of the pan.
4. Transfer the dough into a mixing bowl and beat on low speed until it's 55~60°C. Prepare whole eggs in a separate bowl, beaten thoroughly.
5. Gradually beat in the eggs, a little bit at a time.
6. Based on the basic éclair shape, in a deck oven, bake at top heat 190°C, and bottom heat 190°C for 5 minutes, then reduce the bottom heat to 170°C and continue to bake for 50 minutes. Or, in a convection oven, preheat oven to 160°C, then bake for 30 minutes.

GLUTEN
FREE

2 PATE A CHOUX - GLUTEN FREE

파트 아 슈 - 글루텐 프리

12cm
56ea

DECK OVEN
180°C ···· 180°C - 5min
180°C ···· 170°C - 50min

CONVECTION OVEN
160°C - 30min

ingredients

물 316g	316g Water
우유 362g	362g Milk
버터 338g	338g Butter
설탕 20g	20g Sugar
소금 8g	8g Salt
쌀가루 338g	338g Rice flour
달걀전란 616g	616g Whole eggs
잔탄검 2g	2g Xanthan gum

1
2
3-1
3-2
4
5

Process

1. 냄비에 물, 우유, 버터, 설탕, 소금을 넣은 다음, 버터가 완전히 녹고 액체가 끓을 때까지 가열한다.
2. 불에서 내려 체 친 쌀가루와 잔탄검을 넣고 한 덩어리가 될 때까지 충분히 섞어준다.
3. 다시 불 위에 올려 반죽이 전체적으로 윤기가 나고 냄비 바닥에 얇은 막이 생길 때까지 계속해서 재빠르게 섞어주며 골고루 열을 가한다.
4. 반죽을 믹싱볼에 옮긴 후 저속으로 믹싱해 약 55~60℃로 식힌다. 완전히 풀어 놓은 달걀전란을 조금씩 나누어 가며 혼합한다.
5. 완성된 반죽은 기본 에클레어 모양을 기준으로 데크 오븐의 경우 윗불 180℃/아랫불 180℃로 5분간 굽다가 아랫불을 170℃로 낮춰 50분간 굽는다. 컨벡션 오븐의 경우 160℃로 예열한 오븐에서 30분간 굽는다.

1. In a saucepan, heat water, milk, butter, sugar, and salt, until butter is completely melted and the mixture is boiling.
2. Remove from heat, and mix in sifted rice flour and xanthan gum until it forms one dough.
3. Put the pan back on the heat, and keep stirring quickly to apply heat evenly until the dough is shiny and forms a thin film on the bottom of the pan.
4. Transfer the dough into a mixing bowl and beat on low speed until its 55~60°C. Gradually beat in fully beaten eggs, a little bit at a time.
5. Based on the basic éclair shape, in a deck oven, bake at top heat 180°C, and bottom heat 180°C for 5 minutes, then reduce the bottom heat to 170°C and continue to bake for 50 minutes. Or, in a convection oven, preheat oven to 160°C, then bake for 30 minutes.

CHOUX DOUGH

BASIC 1

2

3

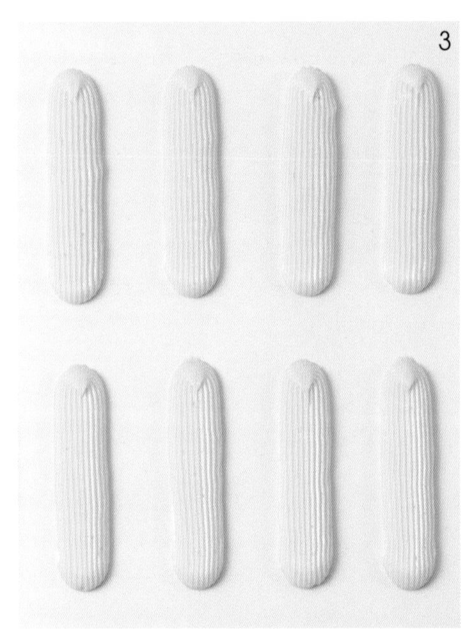

CHOUX A LA CREME 1

2

MOLD 1

2

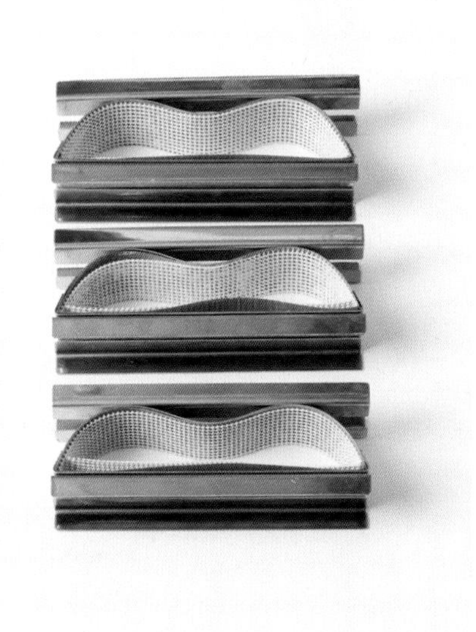

3

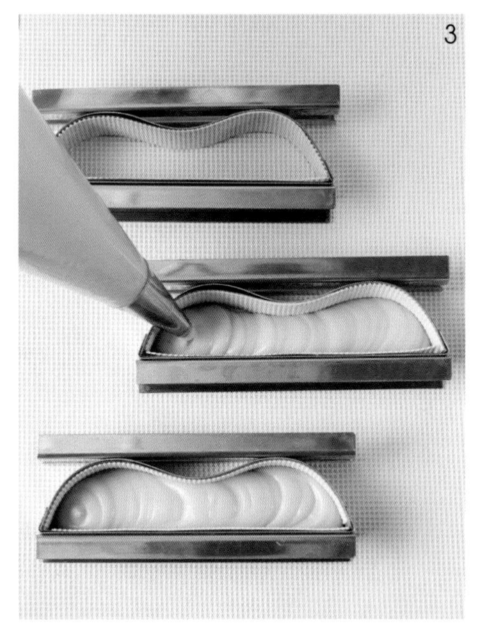

3 DOUGH FORMATION & BAKING

반죽 성형 & 굽기

Process

기본

1. 톱니 모양(PF18번) 깍지를 이용해 12cm 에클레어 모양으로 슈 반죽을 파이핑한다.
2. 체를 이용해 슈거파우더를 얇게 도포한다.
3. 데크 오븐 기준 윗불 190℃/아랫불 190℃로 5분간 굽다가 아랫불을 170℃로 낮춰 50분, 컨벡션 오븐 기준 160℃에서 30분간 굽는다.

슈 아 라 크렘

1. 원형 깍지(지름 1cm)를 이용해 지름 4.8cm 원형으로 슈 반죽을 파이핑한다.
2. 스트로이젤을 올린 후 데크 오븐 기준 윗불 190℃/아랫불 190℃로 5분간 굽다가 아랫불을 170℃로 낮춰 50분, 컨벡션 오븐 기준 160℃에서 30분간 굽는다.

몰드

1. 버터, 붓, 틀 높이와 둘레에 맞춰 재단한 테프론시트와 에어매트, 틀을 준비한다.
2. 에어매트에 버터를 얇게 바른 후 테프론시트 - 에어매트 순서로 틀 안쪽에 둘러준다.
3. 틀 높이의 40% 정도 슈 반죽을 파이핑한 후 에어매트와 무게감 있는 철판을 올려준다. 컨벡션 오븐 기준 180℃에서 25분간 굽다가 150℃로 낮춰 30~35분간 굽는다.

BASIC

1. Pipe the choux paste into a 12cm éclair shape using a petit-four star nozzle (PF18).
2. Dust thin layer of powdered sugar using a sieve.
3. In a deck oven, bake at top heat 190°C, and bottom heat 190°C for 5 minutes, then reduce bottom heat to 170°C and bake for 50 minutes. Or, in a convection oven, bake for 30 minutes at 160°C.

CHOUX A LA CREME

1. Pipe choux paste into 4.8cm diameter rounds using a round tip (1cm diameter).
2. Place streusel on top and bake; in a deck oven, top heat 190°C, and bottom heat 190°C for 5 minutes, then reduce heat to 170°C and bake for 50 minutes, or in a convection oven, 30 minutes at 160°C.

MOLD

1. Prepare butter, brush, Teflon sheet, and air mat cut to the measurement of the mold's height and circumference and the molds.
2. Brush a thin layer of butter on the air mat, then wrap inside the mold in the order of Teflon sheet - air mat.
3. Pipe choux paste in the mold up to 40% of its height, place an air mat on top, and then a heavy baking tray on top of the mat. Bake in a convection oven for 25 minutes at 180°C, then reduce heat to 150°C and continue to bake for 30~35 minutes.

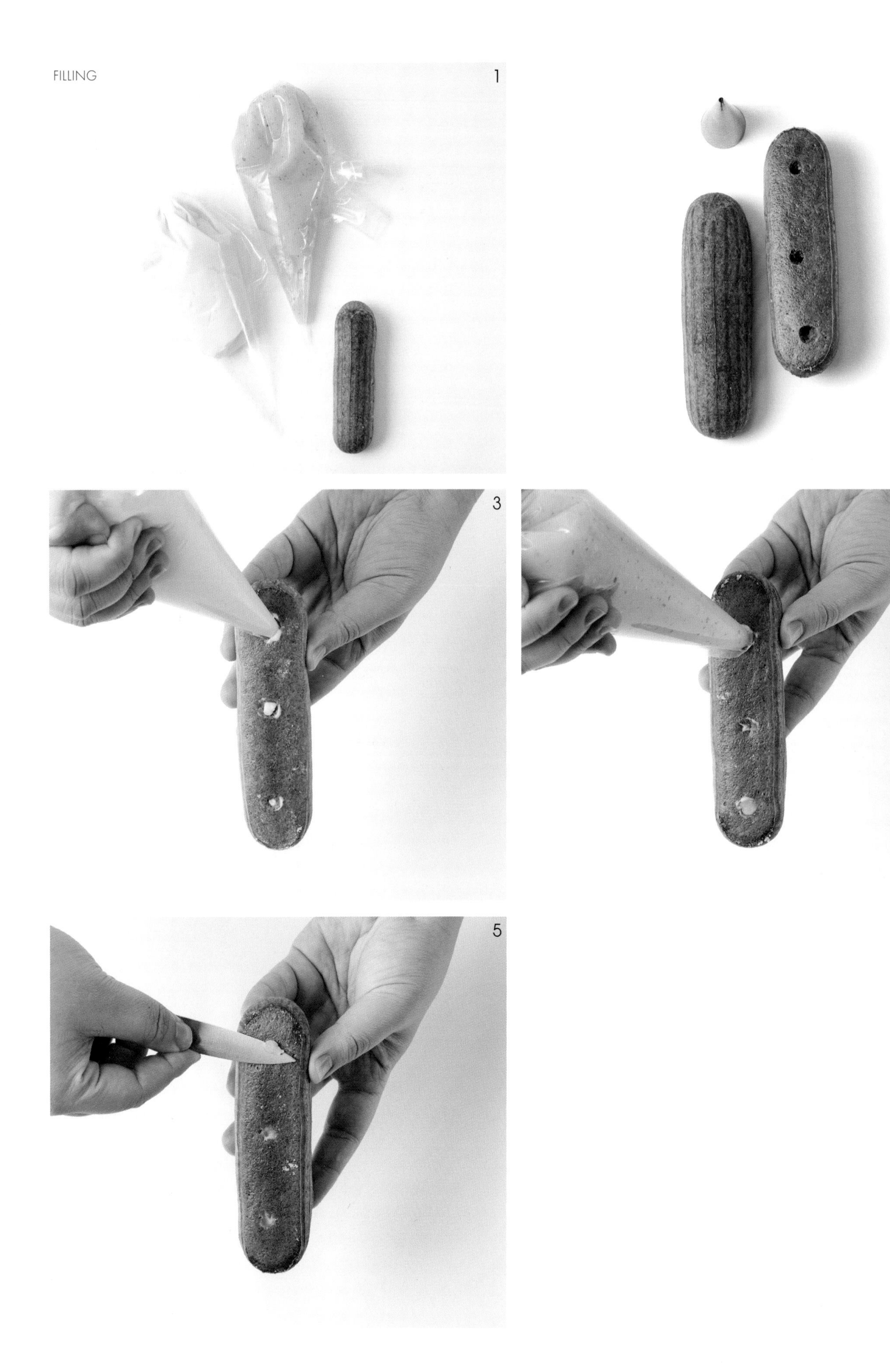
1
2
3
4
5

4 FILLING

크림 충전

Process

1. 구워져 나온 슈와 필링 재료를 준비한다.
2. 작은 원형 깍지를 이용해 슈 바닥에 3개의 필링 주입구를 만들어준다.
3. 크림, 잼 등 슈 안에 채울 필링 재료를 파이핑한다. 필링이 한 가지인 경우 슈가 팽창하는 것이 느껴질 정도로 빈 공간 없이 가득 파이핑한다.
4. 슈에 채워지는 필링이 두 가지 이상인 경우 마지막 필링으로 빈 공간 없이 파이핑한다. 이때 충전물의 비율은 기호에 따라 정할 수 있다.
5. 주입구로 빠져나온 필링은 나이프를 이용해 깔끔하게 정리한다.

1. Prepare baked choux and fillings to use.
2. Make three injection holes on the bottom of the choux using a small round tip.
3. Pipe in fillings such as cream, jam, etc. If only one kind of filling is used, pipe in until you feel the choux expand and there is no empty space.
4. If two or more filling is being piped, completely fill in with the last filling used. The ratio of fillings can be adjusted according to preference.
5. Neatly clean any fillings left out around the injection holes with a knife.

1

2

3

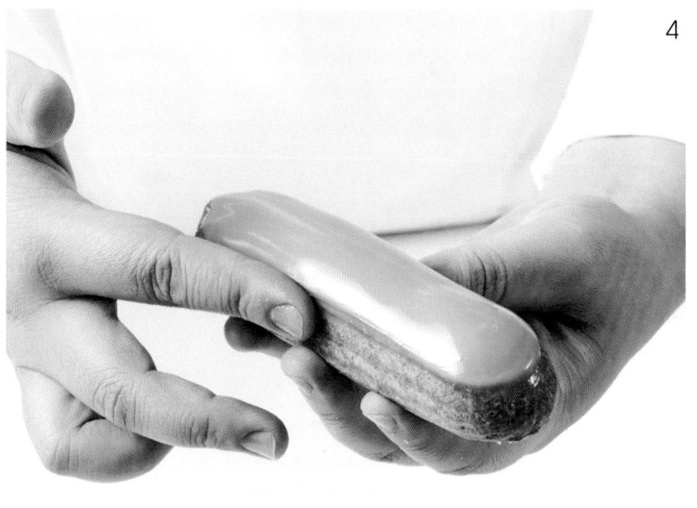
4

5

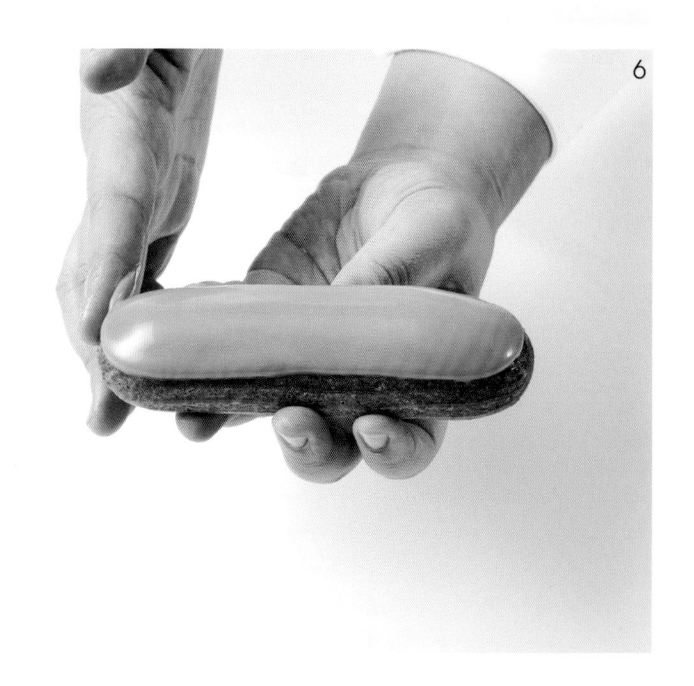
6

5 GLAZING

글레이징

글레이즈의 광택이 살아 있고 울퉁불퉁한 부분이 없으며, 위에서 보았을 때 에클레어 표면이 살짝 보이는 정도로 글레이징하는 것이 완성도가 높습니다.

Glazing quality is considered high when the glaze has a lively sheen without irregularities and can slightly see the surface of the baked éclair when seen straight down from the top.

GELATIN MASS
1
2-1
2-2
3

6 GELATIN MASS

젤라틴매스

응고제의 일종인 젤라틴은 무스나 젤리 등을 굳히는 데 사용되는 재료입니다. 이 책에서는 가루 형태의 젤라틴(200bloom)을 5배의 물과 혼합해 굳힌 후 사용하였습니다.

Gelatine is a kind of coagulant that is used to harden mousse or jelly. For this book, the powder type of gelatin (200 bloom) is mixed with five times of water and hardened to use (gelatin mass).

ingredients

200 Bloom 가루젤라틴 10g
물 50g

10g Powdered gelatin, 200 bloom
50g Water

Process

1. 가루젤라틴과 물을 1:5 비율로 계량한다.
2. 가루젤라틴과 물을 혼합한다.
3. 젤라틴이 굳으면 필요한 만큼 적당한 크기로 잘라 사용한다. 냉장 상태에서 2주, 냉동 상태에서 3개월 동안 보관하며 사용할 수 있다.

1. Scale powdered gelatin and water in the ratio of 1:5.
2. Mix powdered gelatin and water.
3. When the gelatin is set, cut into moderate size to use. It can be stored for about two weeks in a refrigerator and three months in a freezer.

1

7 CHOCOLATE TEMPERING

초콜릿 템퍼링(접종법)

접종법은 녹인 초콜릿에 안정된 결정 상태의 초콜릿을 넣어 템퍼링하는 방법입니다. 여러 가지 템퍼링 방법 중 작업 방식이 간편해 대량 작업 시에도 비교적 손쉽게 템퍼링할 수 있습니다.

The seeding method is a technique of tempering, where stable crystallized chocolate is added to melted chocolate. This method is simple among various methods of tempering, which also makes it relatively easy to work with a large amount of chocolate.

ingredients

커버추어 초콜릿

couverture chocolate

Process

1. 커버추어 초콜릿을 폴리카보네이트 볼에 담고 녹여준다.
(다크초콜릿 55~58℃, 밀크초콜릿 45~48℃, 화이트초콜릿 45~48℃)
2. 녹인 초콜릿 양의 30% 정도의 초콜릿을 1에 넣고 골고루 저어준 후 1분간 그대로 둔다.
3. 핸드블렌더를 이용해 초콜릿 덩어리가 남지 않도록 균일하게 믹싱하며 온도를 낮춰준다.
(다크초콜릿 31~32℃, 밀크초콜릿 29~30℃, 화이트초콜릿 28~29℃)
* 이때 빠른 속도로 장시간 믹싱하면 마찰열에 의해 온도가 과도하게 올라갈 수 있으므로 주의한다.
4. 스패출러 또는 나이프 끝부분에 템퍼링 테스트를 한 후 사용한다.
* 23~24℃를 유지한 작업실에서 5분 이내에 얼룩 없이 매끄럽게 초콜릿이 굳으면 템퍼링이 잘된 상태이다.

1. Melt couverture chocolate in a polycarbonate bowl. (Dark chocolate 55~58°C, milk chocolate 45~48°C, white chocolate 45~48°C)
2. Stir in evenly about 30% of the melted chocolate to 1, and let stand for 1 minute.
3. Reduce the temperature of chocolate using a hand blender so that it's mixed evenly and no chocolate chunks are left.
(Dark chocolate 31~32°C, milk chocolate 29~30°C, white chocolate 28~29°C)
* Keep in mind when mixing at high speed for a long time, the temperature can excessively increase due to heat caused by friction.
4. Test on the tip of a spatula or knife before use.
* If the chocolate hardens smoothly within 5 minutes in a studio that maintained 23~24°C, the tempering is done properly.

1
2
3
4
5
6

8 CORNET

코르네

삼각형 모양으로 준비한 베이킹 페이퍼를 고깔 모양으로 말아준 후 끝부분을 안쪽으로 접어 풀리지 않도록 고정시켜 사용합니다.

Prepare baking paper cut into a triangle, and roll into a cone shape. Fold the end inward to fasten so that it won't undo itself.

Fruits

STRAWBERRY & BASIL ECLAIR

TROPICAL ECLAIR

BLUEBERRY ECLAIR

YUJA & LEMON VERBENA ECLAIR

MOJITO ECLAIR

RASPBERRY KISS ECLAIR

1 STRAWBERRY & BASIL ECLAIR

스트로베리 & 바질 에클레어

ingredients - 15ea

파트 아 슈

BASIC GLUTEN FREE

장식물

딸기
바질

딸기 & 바질 젤

물 53g
딸기 퓌레 125g
레몬주스 27g
아가아가 3.8g
설탕 41g
스위트바질 4g

딸기 & 바질 인서트

딸기 & 바질 젤 226g
딸기 370g

바질 휩 크림

우유 81g
젤라틴매스 12g
화이트초콜릿 72g
(IVOIRE 35%)
생크림 246g
스위트 바질 6g

딸기 소스

딸기 퓌레 133g
레몬주스 27g
카소나드 40g

PATE A CHOUX

BASIC GLUTEN FREE

DECORATION

Strawberries
Basil

STRAWBERRY & BASIL GEL

53g Water
125g Strawberry purée
27g Lemon juice
3.8g Agar-agar
41g Sugar
4g Sweet Basil

STRAWBERRY & BASIL INSERT

226g Strawberry & basil gel
370g Strawberries

BASIL WHIPPED CREAM

81g Milk
12g Gelatin mass
72g White chocolate
(IVOIRE 35%)
246g Fresh cream
6g Sweet basil

STRAWBERRY SAUCE

133g Strawberry purée
27g Lemon juice
40g Cassonade

STRAWBERRY & BASIL GEL
1
2
STRAWBERRY & BASIL INSERT
3
BASIL WHIPPED CREAM
4
5
6
7

Process

딸기 & 바질 젤

1. 냄비에 물, 딸기 퓌레, 레몬주스를 넣고 가열한다. 45℃가 되면 미리 혼합해둔 설탕과 아가아가를 넣고 끓어오를 때까지 가열한 후 볼에 옮겨 냉장고에서 굳힌다.
2. 바질을 넣고 핸드블렌더를 이용해 부드러운 젤 상태가 될 때까지 갈아준다.

딸기 & 바질 인서트

3. 딸기&바질 젤과 사방 3mm 이하로 작게 자른 딸기를 섞어 딸기&바질 인서트를 완성한다.

바질 휩 크림

4. 우유를 끓기 직전까지 가열한 후 젤라틴매스를 혼합한다.
5. 35℃로 녹인 화이트초콜릿에 **4**를 붓고 핸드블렌더로 혼합한다.
6. 35℃로 데운 생크림과 스위트 바질을 넣고 계속해서 핸드블렌더로 혼합한 후 체에 걸러준다.
7. 냉장고에서 12시간 휴지시킨 후 휘핑해 사용한다.

STRAWBERRY & BASIL GEL

1. In a saucepan, heat water, strawberry purée, and lemon juice. When it reaches 45°C, stir in previously mixed sugar and agar-agar, and heat until the mixture boils. Transfer into a bowl and store in the fridge until it sets.
2. Add basil to set base. Use a hand blender to mix until the mixture turns into a soft gel.

STRAWBERRY & BASIL INSERT

3. Cut fresh strawberries into 3mm cubes, then mix with the gel to finish the insert.

BASIL WHIPPED CREAM

4. Heat milk until just before boiling, then stir in gelatin mass.
5. Pour milk mixture(**4**) over white chocolate melted to 35°C. Incorporate with a hand blender.
6. Heat fresh cream to 35°C. Mix in sweet basil using a hand blender and strain.
7. Leave to set in the fridge for 12 hours before whipping.

STRAWBERRY SAUCE

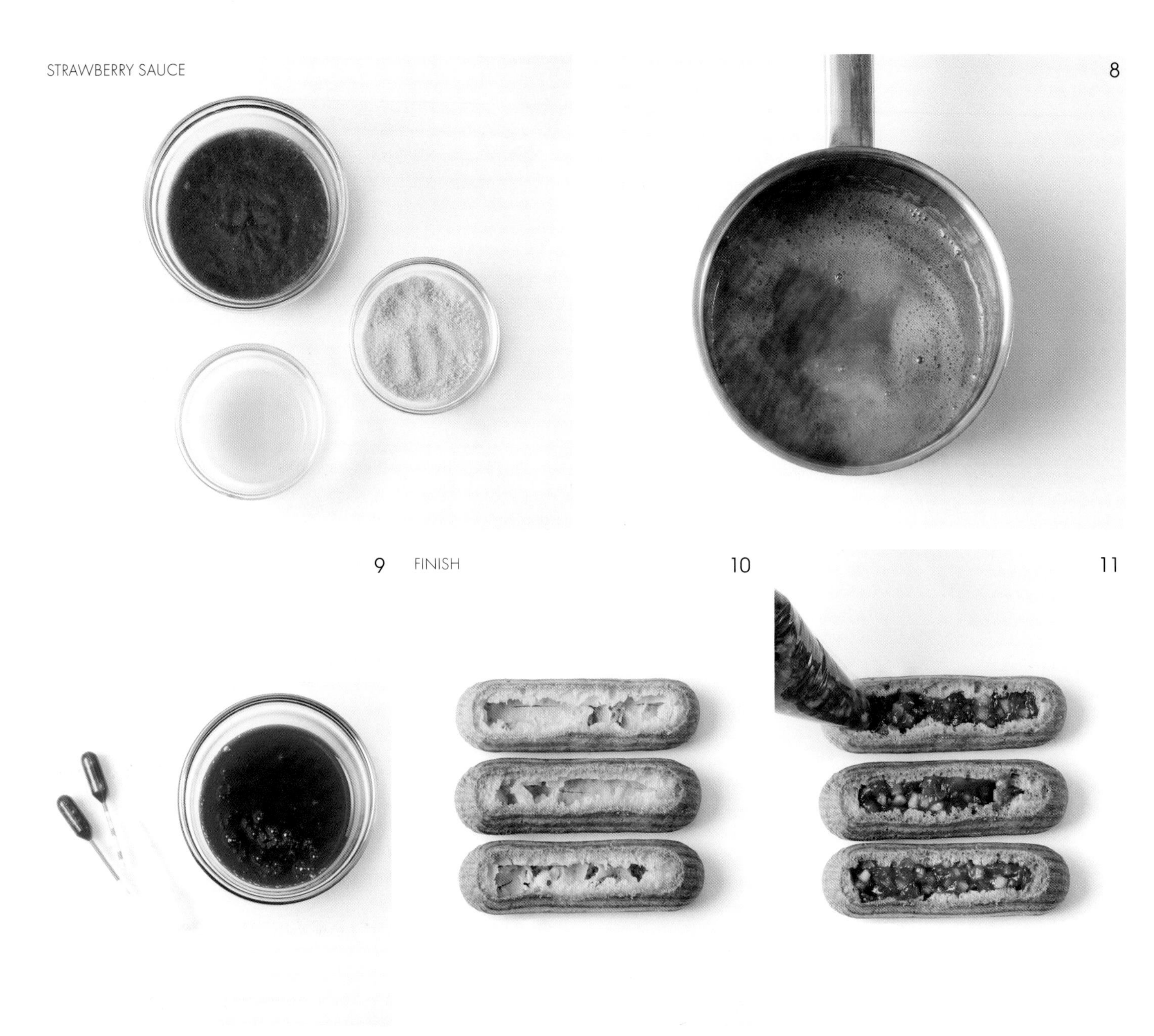

FINISH

10

11

딸기 소스

8. 냄비에 딸기 퓌레, 레몬주스, 카소나드를 넣고 소스 상태로 졸여준다.
9. 완성된 딸기 소스는 차갑게 식힌 후 피펫에 담아준다.

마무리

10. 구워낸 에클레어의 윗부분을 제거한다.
11. 딸기&바질 인서트를 가득 채운다.
12. 톱니 모양(PF12번) 깍지를 이용해 바질 휩 크림을 파이핑한다.
13. 신선한 딸기, 바질잎, 딸기 소스가 담긴 피펫으로 장식해 마무리한다.

STRAWBERRY SAUCE

8. Boil strawberry purée, lemon juice, and cassonade. Reduce to a sauce texture.
9. Cool the strawberry sauce completely. Fill into pipettes.

FINISH

10. Remove the top part of the baked éclairs.
11. Fill with strawberry & basil insert.
12. Pipe basil whipped cream using star nozzle (PF12).
13. Decorate with fresh strawberries, basil leaves, and pipettes filled with strawberry sauce.

2 TROPICAL ECLAIR

트로피칼 에클레어

ingredients - 15ea

파트 아 슈

BASIC

GLUTEN FREE

트로피칼 크림

달걀전란 179g

설탕 123g

망고 퓌레 65g

패션후르츠 퓌레 65g

레몬주스 18g

젤라틴매스 13g

버터 249g

코코넛 휩 크림

코코넛밀크 80g

우유 64g

젤라틴매스 12g

화이트초콜릿 72g
(IVOIRE 35%)

생크림 183g

장식물

코코넛

망고

패션후르츠 씨

허브

PATE A CHOUX

BASIC

GLUTEN FREE

TROPICAL CREAM

179g Whole eggs

123g Sugar

65g Mango purée

65g Passion fruits purée

18g Lemon juice

13g Gelatin mass

249g Butter

COCONUT WHIPPED CREAM

80g Coconut milk

64g Milk

12g Gelatin mass

72g White chocolate
(IVOIRE 35%)

183g Fresh cream

DECORATION

Coconut

Mango

Passion fruit seeds

Herbs

1
2
3
4-1
4-2
5
6-1

Process

트로피칼 크림

1. 볼에 달걀전란과 설탕 1/2을 넣고 혼합한다.
2. 냄비에 망고 퓌레, 패션후르츠 퓌레, 레몬주스, 나머지 설탕을 넣고 45℃로 가열한다.
3. **1**에 **2**를 조금씩 나눠 넣어가며 섞어준다.
4. 다시 냄비에 옮겨 75~78℃로 가열한 후 체로 걸러준다.
5. 젤라틴매스를 혼합한 후 45℃까지 식힌다.
6. 상온 상태의 버터를 넣고 핸드블렌더로 혼합한 후 냉장고에서 12시간 휴지시켜 사용한다.

TROPICAL CREAM

1. Mix whole eggs and 1/2 of the sugar in a bowl.
2. In a saucepan, heat mango purée, passion fruits purée, lemon juice, and remaining sugar to 45°C.
3. Divide the purée mixture(**2**) and gradually mix it into the egg mixture(**1**), a little bit at a time.
4. Place back into a saucepan and heat until it reaches 75~78°C. Strain.
5. Stir in gelatin mass. Cool to 45°C.
6. Incorporate with room temperature butter using a hand blender, and place in the fridge for 12 hours until needed.

COCONUT WHIPPED CREAM
7
8
9
10
COCONUT DECORATION
11
FINISH
12
13
14

코코넛 휩 크림

7. 코코넛밀크와 우유를 끓기 직전까지 가열한 후 젤라틴매스를 혼합한다.
8. 35℃로 녹인 화이트초콜릿에 **7**을 붓고 핸드블렌더로 혼합한다.
9. 35℃로 데운 생크림을 넣고 계속해서 핸드블렌더로 혼합한다.
10. 냉장고에서 12시간 휴지시킨 후 휘핑해 사용한다.

코코넛 장식물

11. 코코넛의 단단한 껍질을 제거하고 필러로 모양내어 깎아 사용한다.

마무리

12. 구워낸 에클레어의 윗부분을 제거한 후 트로피칼 크림을 가득 채운다.
13. 원형 깍지(지름 1cm)를 이용해 코코넛 휩 크림을 파이핑한다.
14. 망고, 허브, 코코넛으로 장식해 마무리한다.

COCONUT WHIPPED CREAM

7. Heat coconut milk and milk until just before boiling. Stir in gelatin mass.
8. Pour the milk mixture(7) over white chocolate melted to 35°C. Mix with a hand blender.
9. Add fresh cream heated to 35°C. Continue to mix using a hand blender.
10. Set in the fridge for 12 hours. Whip to use.

COCONUT DECORATION

11. Remove the hard skin of the coconut. Use a peeler to shave the fruit to use.

FINISH

12. Cut the top part of the baked éclair and completely fill with tropical cream.
13. Use a 1cm round tip to pipe coconut whipped cream.
14. Use fresh mango, herbs, and shaved coconut pieces to decorate.

3 BLUEBERRY ECLAIR

블루베리 에클레어

ingredients - 15ea

파트 아 슈

BASIC GLUTEN FREE

장식물

슈거파우더
블루베리
허브

블루베리 & 레몬 잼

냉동 야생 블루베리 158g
레몬주스 79g
바닐라빈 1/4개
설탕 60g
NH펙틴 3g

프로마주 크림

크림치즈 137g
마스카르포네치즈 39g
연유 33g
설탕 33g
생크림 233g
레몬주스 27g

마스카르포네 휩 크림

우유 65g
젤라틴매스 7.2g
화이트초콜릿 72g
(IVOIRE 35%)
생크림 183g
마스카르포네치즈 78g

PATE A CHOUX

BASIC GLUTEN FREE

DECORATION

Powdered sugar
Blueberries
Herbs

BLUEBERRY & LEMON JAM

158g Frozen wild blueberries
79g Lemon juice
1/4pc Vanilla bean
60g Sugar
3g Pectin NH

FROMAGE CREAM

137g Cream cheese
39g Mascarpone cheese
33g Condensed milk
33g Sugar
233g Fresh cream
27g Lemon juice

MASCARPONE WHIPPED CREAM

65g Milk
7.2g Gelatin mass
72g White chocolate
(IVOIRE 35%)
183g Fresh cream
78g Mascarpone cheese

BLUEBERRY & LEMON JAM
1
2
3
FROMAGE CREAM
4
5
6

Process

블루베리 & 레몬 잼

1. 설탕과 NH펙틴을 섞어준다.
2. 냄비에 냉동 야생 블루베리, 레몬주스, 바닐라빈을 넣고 45℃까지 가열한 후 1을 넣고 혼합한다.
3. 끓어오를 때까지 가열한 다음 볼에 옮겨준다. 완성된 잼은 차갑게 식혀 사용한다.

프로마주 크림

4. 크림치즈와 마스카르포네치즈를 부드럽게 풀어준 후 설탕과 연유를 넣고 섞어준다.
5. 치즈가 덩어리지지 않도록 생크림을 조금씩 나눠 넣어가며 혼합한다.
6. 레몬주스를 넣고 휘핑해 힘 있는 프로마주 크림을 완성한다.

BLUEBERRY & LEMON JAM

1. Mix sugar and pectin NH.
2. In a saucepan, heat frozen wild blueberries, lemon juice, and vanilla bean to 45°C. Whisk in sugar & pectin mixture(1).
3. Continue to heat until the mixture boils. Place in a bowl and cool completely before use.

FROMAGE CREAM

4. Mix cream cheese and mascarpone cheese until soft. Stir in sugar and condensed milk.
5. Gradually mix fresh cream and mix with cheese mixture, making sure to incorporate without any lumps.
6. Add lemon juice and whip to make a firm fromage cream.

7
8-1
8-2
9
10
11-1
11-2

마스카르포네 휩 크림

7. 우유를 끓기 직전까지 가열한 후 젤라틴매스를 혼합한다.
8. 35℃로 녹인 화이트초콜릿에 **7**을 붓고 핸드블렌더로 혼합한다.
9. 35℃로 데운 생크림, 마스카르포네치즈를 넣는다.
10. 계속해서 핸드블렌더로 혼합한다.
11. 냉장고에서 12시간 휴지시킨 후 휘핑해 사용한다.

MASCARPONE WHIPPED CREAM

7. Heat milk until just before boiling. Stir in gelatin mass.
8. Pour milk mixture(7) over white chocolate melted to 35°C. Incorporate with a hand blender.
9. Add fresh cream heated to 35°C, and mascarpone cheese.
10. Continue to mix using a hand blender.
11. Set in the fridge for 12 hours. Whip to use.

FINISH

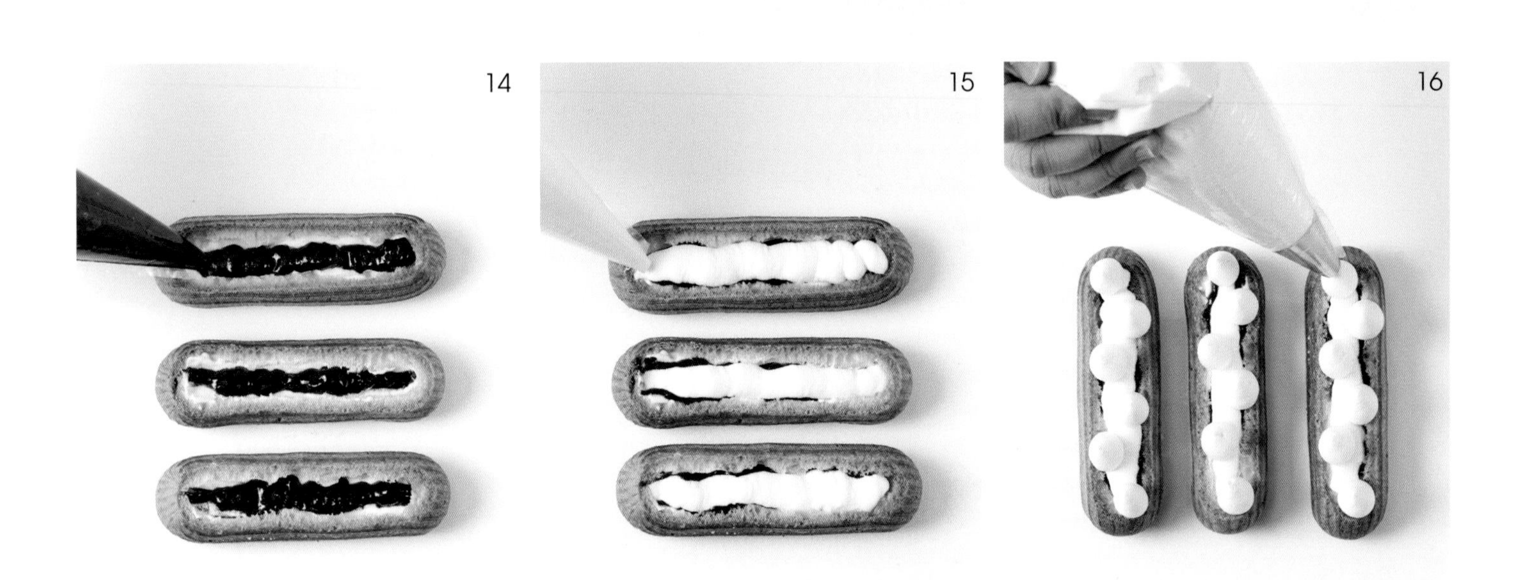

마무리

12. 구워낸 에클레어의 윗부분을 제거한다.
13. 프로마주 크림을 50% 정도 채워준 후 스패출러를 이용해 윗면을 평평하게 다듬어준다.
14. 블루베리&레몬 잼을 적당량 채운다.
15. 다시 프로마주크림을 가득 채운다.
16. 원형 깍지(지름 1cm)를 이용해 마스카르포네 휩 크림을 파이핑한다.
17. 블루베리를 듬뿍 올린 후 슈거파우더를 뿌려준다.
18. 허브로 장식해 마무리한다.

FINISH

12. Cut the top of the baked éclairs.
13. Fill about 50% of the cavity with fromage cream. Flatten the top using a small spatula.
14. Pipe in blueberry & lemon jam.
15. Top with fromage cream.
16. Use 1cm round tip to pipe mascarpone whipped cream.
17. Place fresh blueberries and dust with powdered sugar.
18. Decorate with herbs.

4 YUJA & LEMON VERBENA ECLAIR

유자 & 레몬 버베나 에클레어

ingredients - 15ea

파트 아 슈

BASIC　GLUTEN FREE

장식물

유자초콜릿 (YUZU INSPIRATION)
노란색 천연 식용 색소
레몬 버베나

유자 크림

생크림 320g
젤라틴매스 29.4g
유자초콜릿 247g (YUZU INSPIRATION)
유자주스 128g

유자 & 버베나 젤

물 63g
유자주스 181g
설탕 49g
아가아가 4.5g
레몬 버베나 3g

유자 휩 크림

생크림A 80g
유자주스 60g
젤라틴매스 14.4g
화이트초콜릿 72g (IVOIRE 35%)
생크림B 185g

글레이즈

생크림 183g
물엿 73g
젤라틴매스 42g
화이트초콜릿 231g (OPALYS 33%)
화이트 코팅초콜릿 207g
노란색 천연 식용 색소 적당량

PATE A CHOUX

BASIC　GLUTEN FREE

DECORATION

Yuja chocolate (YUZU INSPIRATION)
Yellow food color (natural)
Lemon Verbena

YUJA CREAM

320g Fresh cream
29.4g Gelatin mass
247g Yuja chocolate (YUZU INSPIRATION)
128g Yuja juice

YUJA & LEMON VERBENA GEL

63g Water
181g Yuja juice
49g Sugar
4.5g Agar-agar
3g Lemon Verbena

YUJA WHIPPED CREAM

80g Fresh cream A
60g Yuja juice
14.4g Gelatin mass
72g White chocolate (IVOIRE 35%)
185g Fresh cream B

GLAZE

183g Fresh cream
73g Corn syrup
42g Gelatin mass
231g White chocolate (OPALYS 33%)
207g White compound chocolate
Yellow food coloring (natural) QS

YUJA CREAM
1
2
3
YUJA & LEMON VERBENA GEL
4
5
6-1
6-2

Process

유자 크림

1. 생크림을 45℃로 데운 후 녹인 젤라틴매스를 혼합한다.
2. 40℃로 녹인 유자초콜릿에 1을 조금씩 넣어가며 섞어준 다음 핸드블렌더로 혼합한다.
3. 30℃로 데운 유자주스를 넣고 계속해서 핸드블렌더로 혼합한 후 냉장고에서 12시간 휴지시켜 사용한다.

유자 & 레몬 버베나 젤

4. 냄비에 물과 유자주스를 넣고 가열한다. 45℃가 되면 미리 혼합해둔 설탕과 아가아가를 넣고 끓어오를 때까지 가열한다.
5. 볼에 옮겨준 후 냉장고에서 굳힌다.
6. 레몬 버베나 잎을 넣고 핸드블렌더를 이용해 부드러운 젤 상태가 될 때까지 갈아준다.

YUJA CREAM

1. Heat fresh cream to 45°C, and stir in gelatin mass.
2. Gradually mix cream mixture(1) with Yuja chocolate melted to 40°C using a hand blender.
3. Warm yuja juice to 30°C and continue to mix with a hand blender. Rest in the fridge for 12 hours before use.

YUJA & LEMON VERBENA GEL

4. In a saucepan, heat water and yuja juice. When it reaches 45°C, add previously mixed sugar and agar-agar mixture and continue to heat until the mixture boils.
5. Place the mixture into a bowl. Set in the fridge.
6. Incorporate lemon verbena leaves and the set base with a hand blender until the mixture turns into a soft gel.

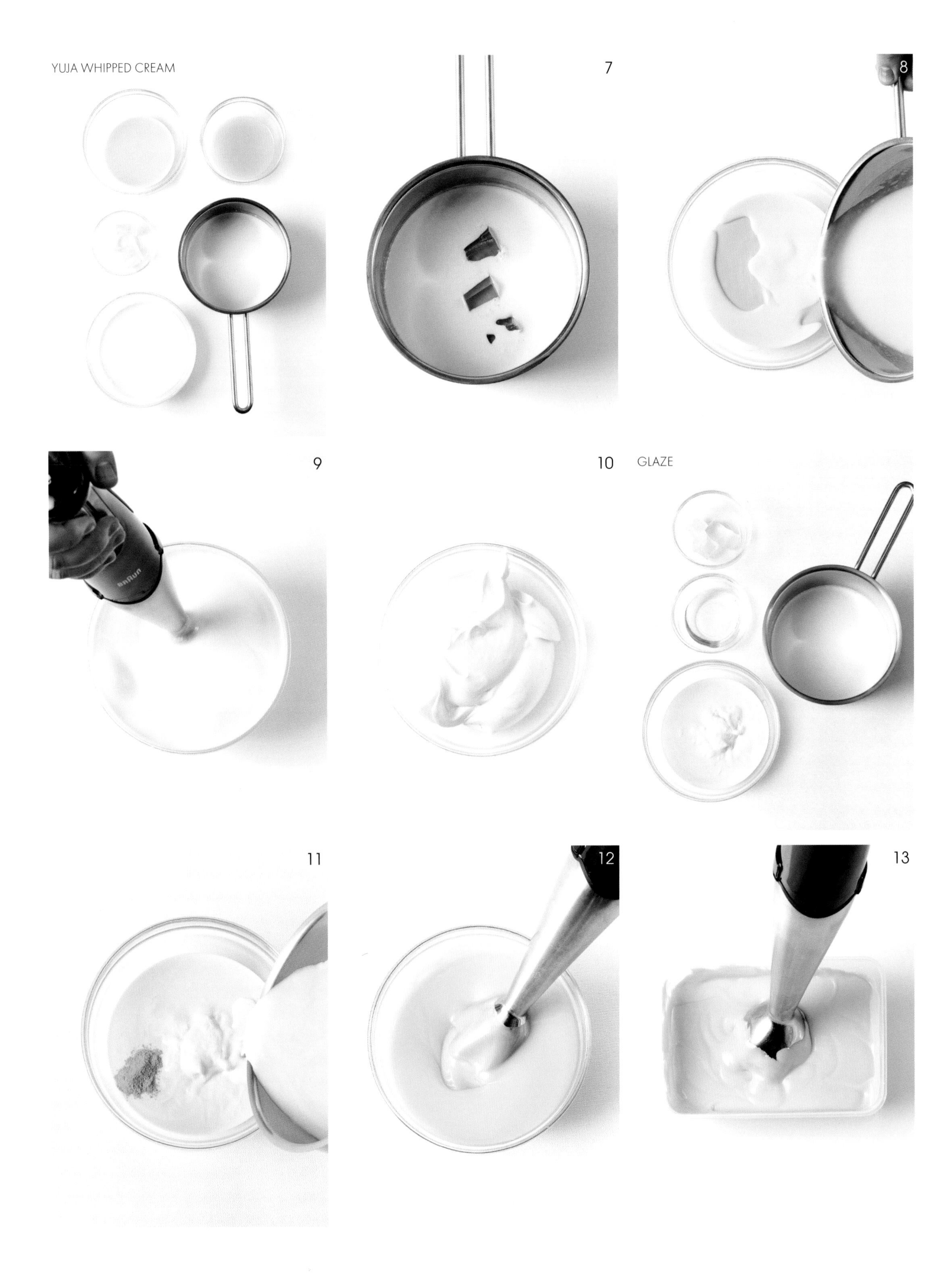
YUJA WHIPPED CREAM
7
8
9
10
GLAZE
11
12
13

유자 휩 크림

7. 끓기 직전까지 가열한 생크림A에 젤라틴매스를 혼합한다.
8. 35℃로 녹인 화이트초콜릿에 7을 붓고 핸드블렌더로 혼합한다.
9. 35℃로 데운 생크림B와 유자주스를 순서대로 넣고 계속해서 핸드블렌더로 혼합한다.
10. 냉장고에서 12시간 휴지시킨 후 휘핑해 사용한다.

글레이즈

11. 끓기 직전까지 가열한 생크림과 물엿에 젤라틴매스를 넣어 녹인 후, 35℃로 녹인 화이트초콜릿과 코팅초콜릿에 부어준다. 이때 노란색 천연 식용 색소를 첨가한다.
12. 핸드블렌더로 혼합한 후 냉장고에서 12시간 세팅시킨다.
13. 사용할 때는 다시 30℃로 온도를 맞춘 후 핸드블렌더로 균일하게 믹싱해 사용한다.

YUJA WHIPPED CREAM

7. Heat milk until just before boiling. Stir in gelatin mass.
8. Pour over white chocolate melted to 35°C. Incorporate using a hand blender.
9. Using a hand blender, continue to mix with fresh cream B heated to 35°C, and then add yuja juice. Blend until incorporated.
10. Set in the fridge for 12 hours. Whip to use.

GLAZE

11. Heat fresh cream and corn syrup until just before boiling. Stir in gelatin mass. Add to white chocolate melted to 35°C and compound chocolate. Add natural yellow food coloring at this stage.
12. Mix with a hand blender. Set in the fridge for 12 hours.
13. To use, reheat the glaze to 30°C, and mix thoroughly using a hand blender.

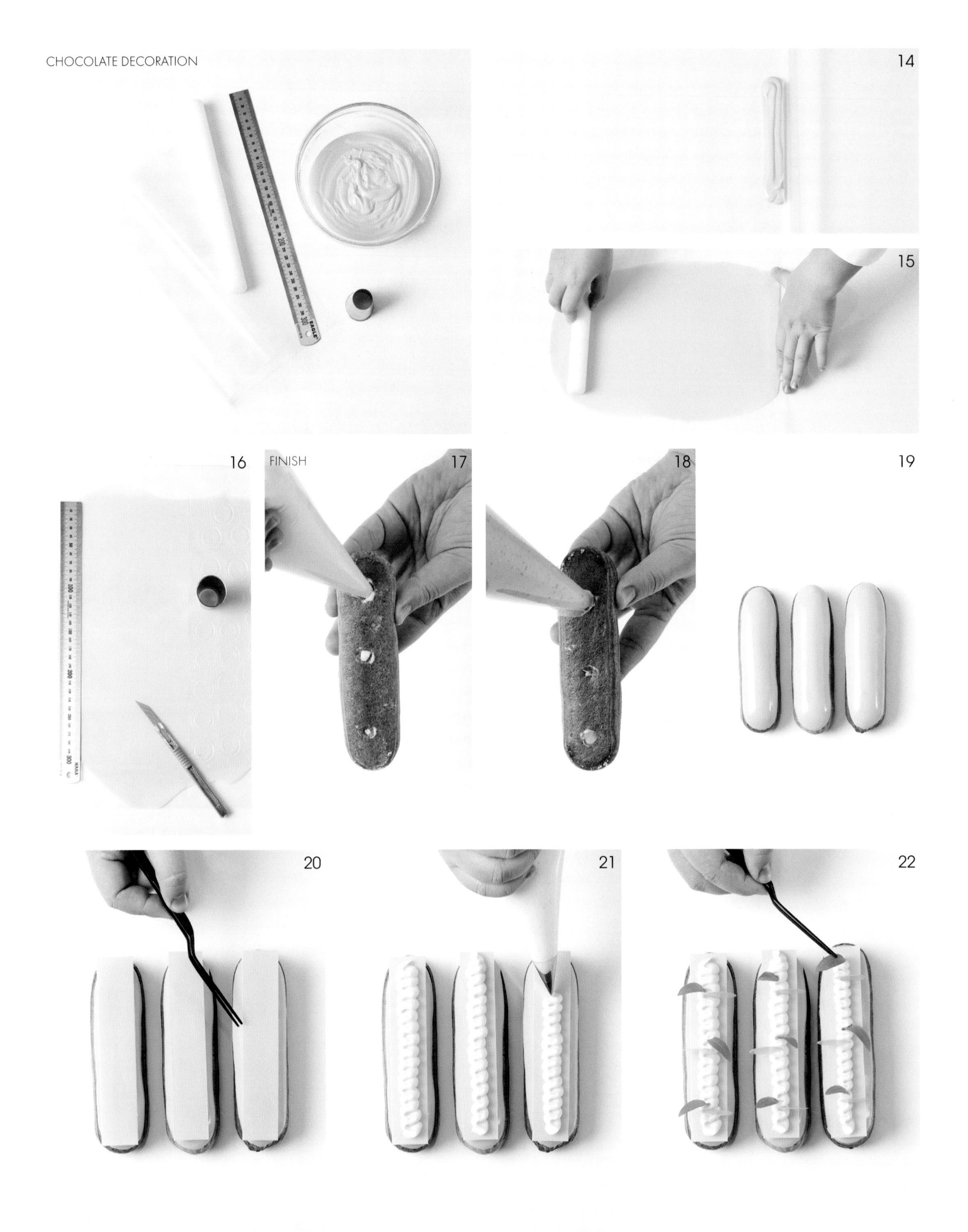
CHOCOLATE DECORATION
14
15
16
FINISH
17
18
19
20
21
22

초콜릿 장식물

14. 두 장의 투명 필름 사이에 노란색 천연 식용 색소를 섞어 템퍼링한 유자초콜릿을 적당량 부어준다.
15. 필름을 덮고 밀대를 이용해 균일한 두께가 되도록 밀어편다.
16. 13×2.5cm 직사각형, 지름 2cm 원형으로 커팅한다.

마무리

17. 구워낸 에클레어 바닥 부분에 작은 원형 깍지를 이용해 3개의 크림 주입구를 만든 후 유자 크림을 90% 정도 채운다.
18. 남은 공간에 유자&레몬 버베나 젤을 채운 후 주입구를 깔끔하게 정리한다.
19. 에클레어 윗면을 글레이징한다.
20. 13×2.5cm 직사각형으로 재단한 초콜릿 장식물을 올려준다.
21. 꽃잎 모양(104번) 깍지를 이용해 유자 휩 크림을 파이핑한다.
22. 원형 초콜릿 디스크와 레몬 버베나 잎으로 장식해 마무리한다.

CHOCOLATE DECORATION

14. Place tempered yuja chocolate mixed with natural yellow food coloring between two sheets of plastic film.
15. Roll into even thickness using a rolling pin.
16. Cut into two sizes; (1) 13 x 2.5cm rectangles, (2) 2cm diameter rounds.

FINISH

17. Using a small round piping tip, make three holes on the bottom of the baked éclairs. Fill with yuja cream about 90% of the cavity.
18. Pipe in yuja & verbena gel in the remaining space, and scrape off the excess.
19. Glaze the top of the éclairs.
20. Place rectangular chocolate piece cut into 13 x 2.5cm on top.
21. Pipe yuja whipped cream using a petal tip (#104).
22. Decorate with round chocolate disks and lemon verbena leaves.

5 MOJITO ECLAIR

모히토 에클레어

ingredients - 15ea

파트 아 슈

BASIC GLUTEN FREE

장식물

라임제스트

라임 크림

달걀전란 176g
설탕 127g
라임주스 141g
젤라틴매스 12.6g
버터 255g

라임 & 애플민트 젤

물 63g
라임주스 181g
설탕 49g
아가아가 4.5g
애플민트 3g

글레이즈

생크림 183g
물엿 73g
젤라틴매스 42g
화이트초콜릿 231g
(OPALYS 33%)
화이트 코팅초콜릿 207g
애플민트 2g
노란색 천연 식용 색소 적당량
녹색 천연 식용 색소 적당량

PATE A CHOUX

BASIC GLUTEN FREE

DECORATION

Lime zest

LIME CREAM

176g Whole eggs
127g Sugar
141g Lime juice
12.6g Gelatin mass
255g Butter

LIME & APPLEMINT GEL

63g Water
181g Lime juice
49g Sugar
4.5g Agar-agar
3g Applemint

GLAZE

183g Fresh cream
73g Corn syrup
42g Gelatin mass
231g White chocolate
(OPALYS 33%)
207g White compound chocolate
2g Applemint
Yellow food coloring (natural) QS
Green food coloring (natural) QS

LIME CREAM
1
2
3
4
5
LIME & APPLEMINT GEL
6
7
8

Process

라임 크림

1. 볼에 달걀전란과 설탕 1/2을 넣고 섞어준다. 냄비에는 라임주스와 나머지 설탕을 넣고 45℃로 가열한다.
2. 1의 달걀 믹스처에 가열한 라임주스를 조금씩 나눠 넣어가며 섞어준다.
3. 2를 다시 냄비에 옮겨 75~78℃로 가열한 후 체에 걸러준다.
4. 젤라틴매스를 혼합한 후 45℃까지 식힌다.
5. 상온 상태의 버터를 넣고 핸드블렌더로 혼합한 후 냉장고에서 12시간 휴지시켜 사용한다.

라임 & 애플민트 젤

6. 냄비에 물과 라임주스를 넣고 가열한다. 45℃가 되면 미리 혼합해둔 설탕과 아가아가를 넣고 끓어오를 때까지 가열한다.
7. 볼에 옮겨준 후 냉장고에서 굳힌다.
8. 애플민트를 넣고 핸드블렌더로 부드러운 젤 상태가 될 때까지 갈아준다.

LIME CREAM

1. In a bowl, mix whole eggs and 1/2 of sugar. In a saucepan, heat lime juice and remaining sugar to 45°C.
2. Gradually mix in heated lime juice mixture into the egg mixture(1), a little bit at a time.
3. Place the mixture back into the saucepan, and cook until the temparature reaches 75~78°C. Strain.
4. Add gelatin mass and mix. Cool down to 45°C.
5. Mix with room temperature butter using a hand blender. Place in the fridge for 12 hours before use.

LIME & APPLEMINT GEL

6. Heat water and lime juice. When the temperature reaches 45°C, stir in previously mixed sugar and agar-agar and boil.
7. Place into a bowl and set in the fridge.
8. Blend the set base with applemint until the mixture turns into a soft gel.

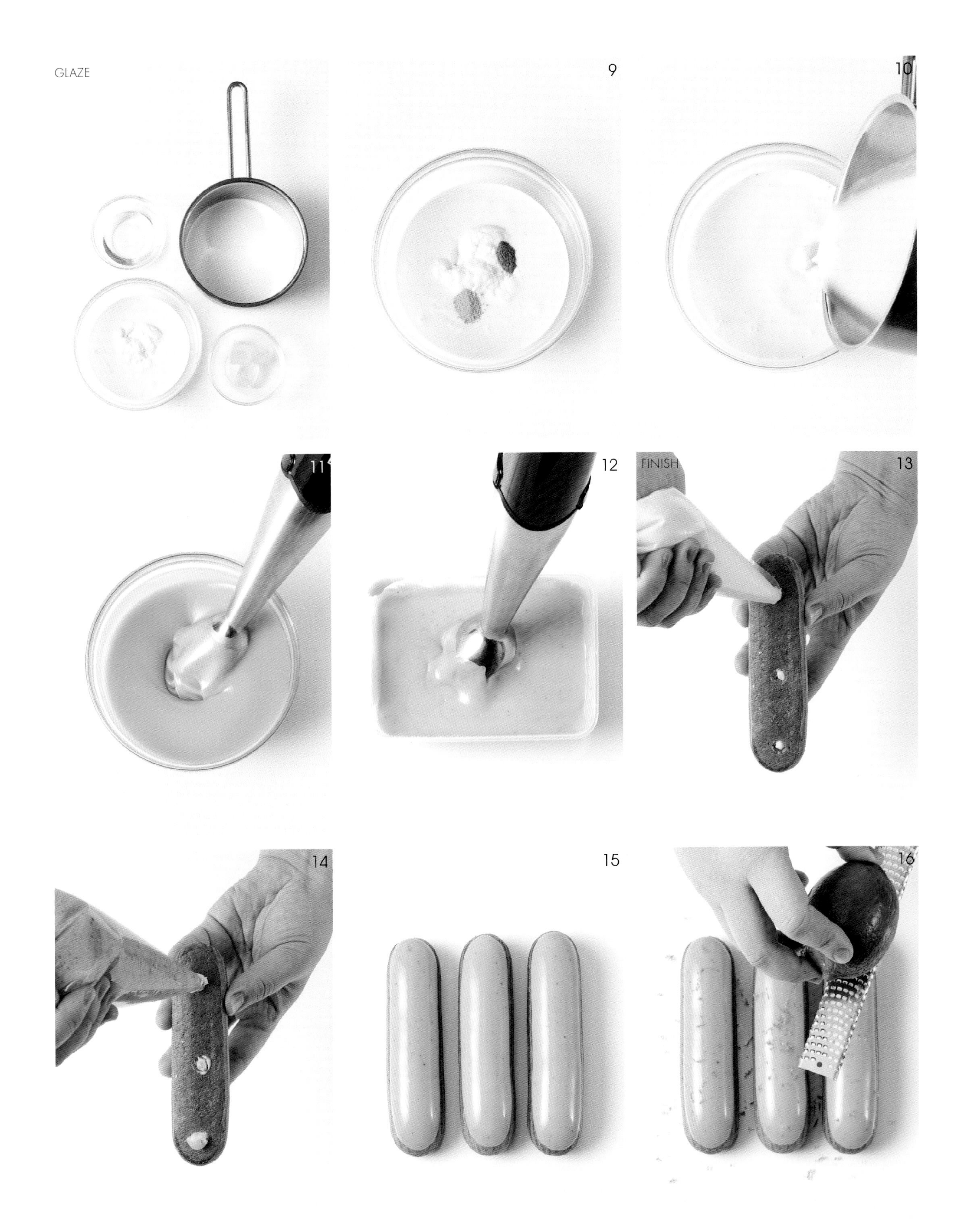
GLAZE
9
10
11
12
FINISH
13
14
15
16

글레이즈

9. 35℃로 녹인 화이트초콜릿과 코팅초콜릿에 노란색과 녹색 천연 식용 색소를 첨가한다.
10. 냄비에 생크림과 물엿을 넣고 끓기 직전까지 가열한 후 젤라틴매스를 넣어 섞고 **9**에 부어준다.
11. 핸드블렌더로 혼합한 후 냉장고에서 12시간 세팅시킨다.
12. 사용할 때는 다시 30℃로 온도를 맞춘 다음 애플민트를 넣고 핸드블렌더로 균일하게 믹싱해 사용한다.

마무리

13. 구워낸 에클레어 바닥 부분에 작은 원형 깍지를 이용해 3개의 크림 주입구를 만든 후 라임 크림을 90% 정도 채운다.
14. 남은 공간에 라임&애플민트 젤을 가득 채우고 주입구를 깔끔하게 정리한다.
15. 에클레어 윗면을 글레이징한다.
16. 라임제스트로 장식해 마무리한다.

GLAZE

9. Mix natural yellow and green food coloring with white chocolate melted to 35°C and compound chocolate.
10. Heat fresh cream and corn syrup until just before boiling, then add gelatin mass. Pour over the chocolate mixture(**9**).
11. Incorporate with a hand blender. Set in the fridge for 12 hours.
12. To use, warm the glaze to 30°C and mix with a hand blender thoroughly.

FINISH

13. Using a small round piping tip, make three holes on the bottom of of the baked éclairs. Fill with yuja cream about 90% of the cavity.
14. Pipe in lime & applemint gel in the remaining space, and scrape off the excess around the holes.
15. Glaze the top of the filled éclairs.
16. Decorate with lime zest.

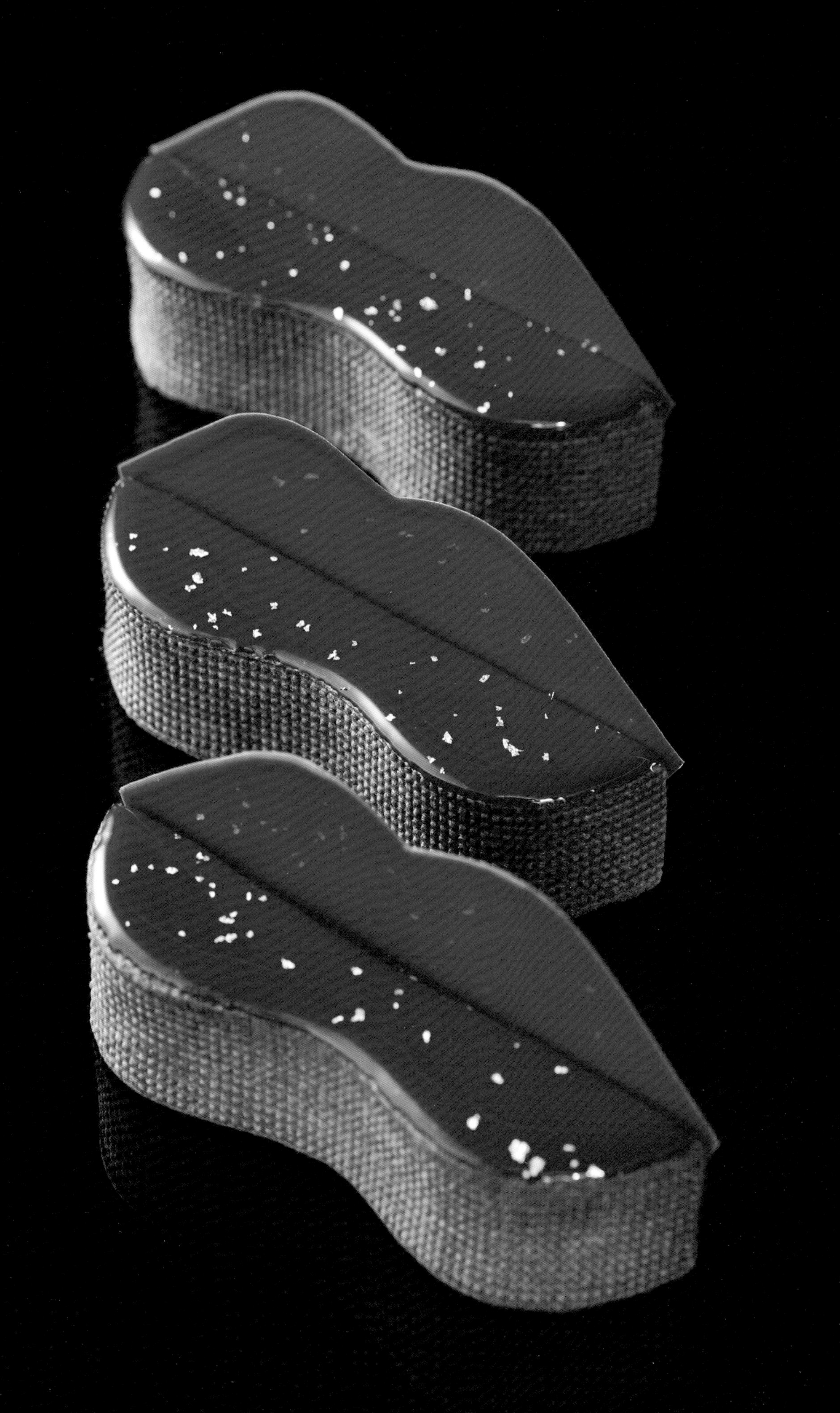

6 RASPBERRY KISS ECLAIR

라즈베리 키스 에클레어

ingredients - 15ea

파트 아 슈

BASIC GLUTEN FREE

장식물

프람보아즈초콜릿
(FRAMBOISE INSPIRATION)

식용금박

라즈베리 크림

생크림 286g

젤라틴매스 25.2g

프람보아즈초콜릿 189g
(FRAMBOISE INSPIRATION)

물 81g

라즈베리 퓌레 141g

라즈베리 잼

냉동 라즈베리 145g

라즈베리 퓌레 145g

설탕 80g

NH펙틴 4g

레몬주스 13g

키르쉬 13g

글레이즈

물 75g

설탕 150g

물엿 150g

연유 100g

젤라틴매스 60g

화이트초콜릿 120g
(OPALYS 33%)

다크초콜릿 30g
(CARAIBE 66%)

붉은색 천연 식용 색소 2.5g

PATE A CHOUX

BASIC GLUTEN FREE

DECORATION

Framboise Chocolate
(FRAMBOISE INSPIRATION)

Edible gold flakes

RASPBERRY CREAM

286g Fresh cream

25.2g Gelatin mass

189g Framboise Chocolate
(FRAMBOISE INSPIRATION)

81g Water

141g Raspberry purée

RASPBERRY JAM

145g Frozen raspberries

145g Raspberry purée

80g Sugar

4g Pectin NH

13g Lemon juice

13g Kirsch

GLAZE

75g Water

150g Sugar

150g Corn syrup

100g Condensed milk

60g Gelatin mass

120g White chocolate
(OPALYS 33%)

30g Dark chocolate
(CARAIBE 66%)

2.5g Red food coloring (natural)

RASPBERRY CREAM
1
2
3
4
RASPBERRY JAM
5
6
7

Process

라즈베리 크림

1. 생크림을 45℃로 데운 후 녹인 젤라틴매스를 혼합한다.
2. 40℃로 녹인 프람보아즈초콜릿에 **1**을 조금씩 나눠 넣어가며 섞어준 다음 핸드블렌더로 혼합한다.
3. 라즈베리 퓌레와 물을 30℃로 데운 후 **2**에 부어준다.
4. 계속해서 핸드블렌더로 혼합한 후 냉장고에서 12시간 휴지시켜 사용한다.

라즈베리 잼

5. 설탕과 NH펙틴을 혼합한다.
6. 냄비에 냉동 라즈베리와 라즈베리 퓌레를 넣고 가열한다. 45℃가 되면 **5**를 넣고 끓어오를 때까지 가열한다.
7. 불에서 내린 후 레몬주스와 키르쉬를 순서대로 넣고 혼합한다. 완성된 라즈베리 잼은 차갑게 식힌 후 사용한다.

RASPBERRY CREAM

1. Warm fresh cream to 45°C. Stir in gelatin mass.
2. Gradually add cream mixture(1) to Framboise chocolate melted to 40°C, and incorporate using a hand blender.
3. Warm raspberry purée and water to 30°C, and add to chocolate mixture(2).
4. Continue to mix with a hand blender. Rest in the fridge for 12 hours before use.

RASPBERRY JAM

5. Mix sugar and pectin NH.
6. Heat frozen raspberries and rasberry purée in a saucepan. When the temparature reaches 45°C, stir in the sugar mixture(5) and bring it to a boil.
7. Remove from heat, and add lemon juice and Kirsch in order. Cool finished raspberry jam completely before use.

GLAZE
8
9
10
11
12
CHOCOLATE DECORATION
13
14
15

글레이즈

8. 35℃로 녹인 화이트초콜릿과 다크초콜릿에 붉은색 천연 식용 색소를 첨가한다.
9. 냄비에 물과 설탕을 넣고 가열한다.
10. 시럽 상태가 되면 물엿을 넣고 103℃까지 가열한 후 불에서 내려준다.
11. 연유와 젤라틴매스를 순서대로 넣어 섞고 **8**에 부어준 후 핸드블렌더로 혼합한다. 냉장고에서 12시간 세팅시킨다.
12. 사용할 때는 다시 30℃로 온도를 맞춘 다음 핸드블렌더로 균일하게 믹싱해 사용한다.

초콜릿 장식물

13. 두 장의 투명 필름 사이에 템퍼링한 프람보아즈초콜릿을 적당량 부어준다.
14. 필름을 덮고 밀대를 이용해 균일한 두께가 되도록 밀어편다.
15. 입술 모양 커터로 커팅한다.

GLAZE

8. Mix white chocolate melted to 35°C and natural red food coloring.
9. In a saucepan, boil water and sugar together.
10. When it turns into syrup, add corn syrup and continue to boil until the temperature reaches 103°C. Remove from heat.
11. Add condensed milk, then gelatin mass in order. Mix into the chocolate mixture(**8**) using a hand blender. Set in the fridge for 12 hours.
12. To use, warm the glaze to 30°C and mix with a hand blender thoroughly.

CHOCOLATE DECORATION

13. Place a small amount of tempered Framboise chocolate between two sheets of plastic film.
14. Roll out to even thickness using a rolling pin.
15. Cut out disks using lip-shaped cutter.

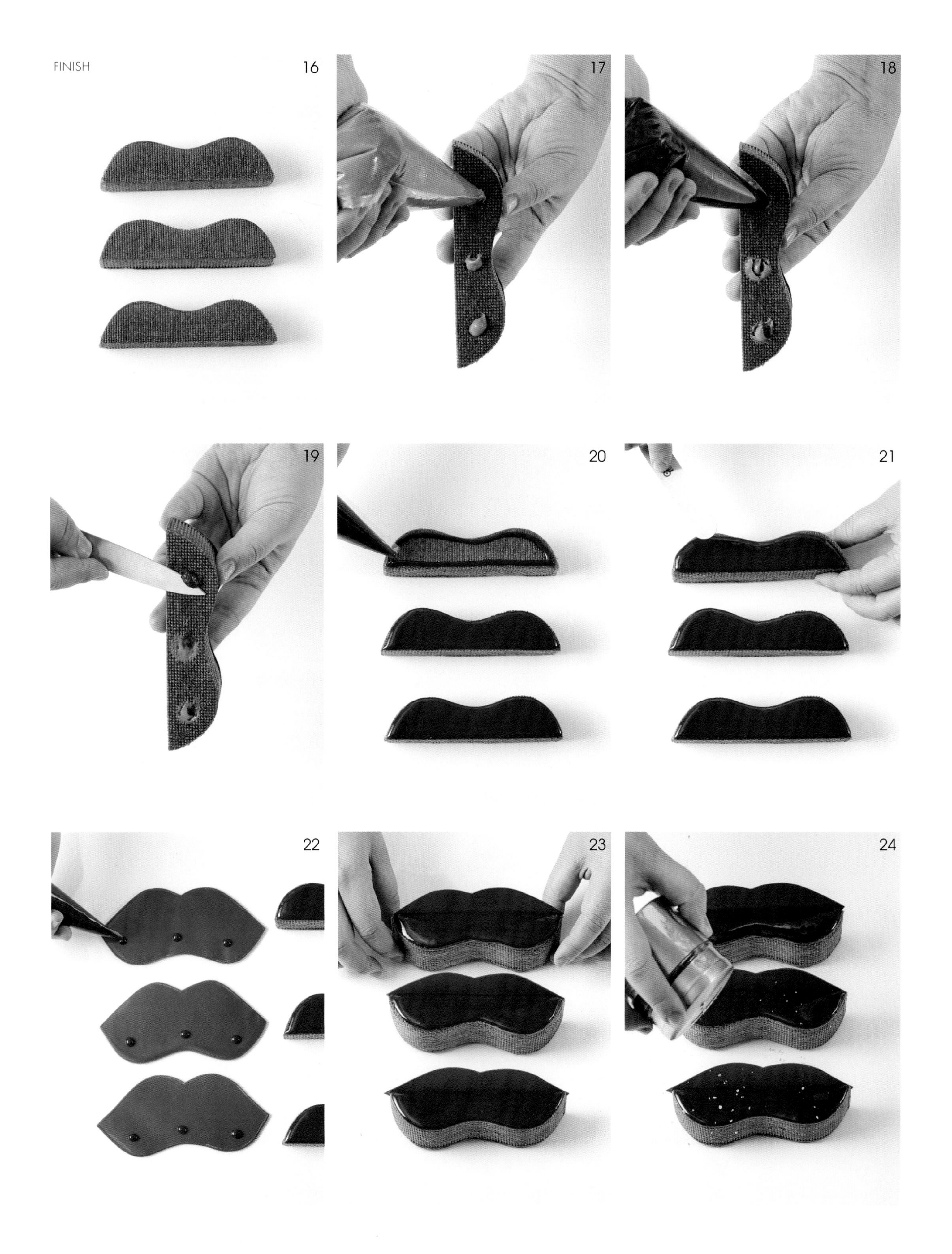
FINISH
16
17
18
19
20
21
22
23
24

마무리

16. 구워낸 에클레어 바닥 부분에 작은 원형 깍지를 이용해 3개의 크림 주입구를 만든다. (47p)
17. 라즈베리 크림을 90% 정도 채운다.
18. 남은 공간에 라즈베리 잼을 가득 채운다.
19. 주입구를 깔끔하게 정리한다.
20. 짤주머니를 이용해 에클레어 윗면을 글레이징한다. 입술 모양을 따라 가장자리를 먼저 그려준 후 안쪽을 채운다.
21. 팔레트나이프를 이용해 가장자리를 정리한다.
22. 초콜릿 장식물에 여분의 글레이즈를 파이핑한다.
23. 초콜릿 장식물을 에클레어 뒷면에 고정시켜 입술 모양을 완성한다.
24. 식용금박을 뿌려 마무리한다.

FINISH

16. Using a small round piping tip, make three holes on the bottom of the baked éclairs. (47p)
17. Fill with raspberry cream about 90% of the cavity.
18. Pipe in the raspberry jam in the remaining space.
19. Scrape off the excess around the holes.
20. Glaze the top of the éclairs using a piping bag. Draw a thin line around the edge, then fill in the rest.
21. Use a palette knife to clean up around the edges.
22. Pipe dots of remaining glaze on chocolate decoration piece.
23. Fix the chocolate piece on the back of the éclair to complete the lip shape.
24. Sprinkle edible gold flakes to finish.

Nut & Chocolate

MONT BLANC ECLAIR

FORET NOIRE ECLAIR

PISTACHIO ECLAIR

BLACK SESAME ECLAIR

BITTER CHOCOLATE ECLAIR

HAZELNUT CHOCOLATE BAR

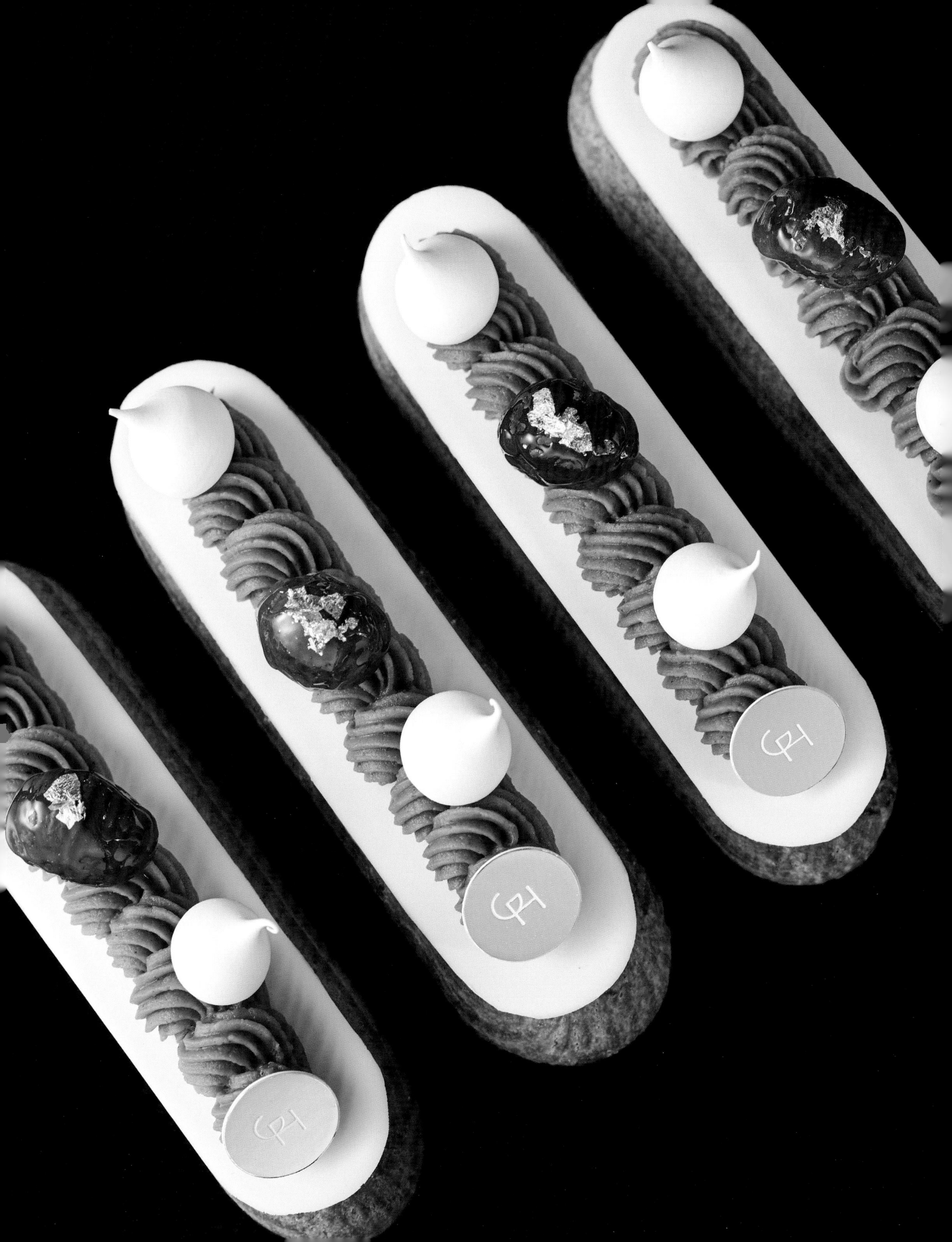

7 MONT BLANC ECLAIR

몽블랑 에클레어

ingredients - 15ea

파트 아 슈

BASIC GLUTEN FREE

체스트넛 크림

달걀노른자 39g
설탕 21g
생크림 54g
우유 54g
젤라틴매스 8.4g
마스카르포네치즈 242g
밤 페이스트 174g
크렘 파티시에(259p) 116g

카시스 & 레몬 잼

카시스 퓌레 158g
레몬주스 79g
설탕 60g
NH펙틴 3g

체스트넛 토핑 크림

밤 페이스트 224g
밤 퓌레 224g
럼 또는 시럽 적당량

스위스 머랭 쿠키*

설탕 100g
달걀흰자 50g

장식물

스위스 머랭 쿠키*
보늬밤
식용금박

PATE A CHOUX

BASIC GLUTEN FREE

CHESTNUT CREAM

39g Egg yolks
21g Sugar
54g Fresh cream
54g Milk
8.4g Gelatin mass
242g Mascarpone cheese
174g Chestnut paste
116g Crème patissiere(259p)

CASSIS & LEMON JAM

158g Cassis purée
79g Lemon juice
60g Sugar
3g Pectin NH

CHESTNUT TOPPING CREAM

224g Chestnut paste
224g Chestnut purée
Rum or syrup QS

SWISS MERINGUE COOKIES*

100g Sugar
50g Egg whites

DECORATION

Swiss meringue cookies*
Whole chestnut with skin in syrup
Edible gold leaves

CHESTNUT CREAM
1
2
3
4-1
4-2
5
6
7

Process

체스트넛 크림

1. 달걀노른자와 설탕 1/2을 혼합한다.
2. 냄비에 생크림, 우유, 나머지 설탕을 넣고 45℃로 가열한다.
3. 1에 2를 조금씩 나눠 넣으면서 섞는다.
4. 냄비에 옮겨 83~85℃로 가열해 크렘 앙글레이즈를 만든다.
5. 체에 내려 볼에 옮겨준 후 젤라틴매스를 혼합한다.
6. 마스카르포네치즈, 밤 페이스트, 크렘 파티시에를 넣고 핸드블렌더로 혼합한다.
7. 냉장고에서 12시간 휴지시킨 후 사용한다.

CHESTNUT CREAM

1. Mix egg yolks and 1/2 of sugar.
2. In a saucepan, heat fresh cream, milk, and remaining sugar to 45°C.
3. Gradually add cream mixture(2) to egg yolk mixture(1), a little bit at a time. Mix well to incorporate.
4. Pour back into the saucepan and heat until the temperature reaches 83~85°C to make crème anglais.
5. Strain the crème into a bowl and mix with gelatin mass.
6. Add mascarpone cheese, chestnut paste, and crème patissiere and mix using a hand blender.
7. Rest in the fridge for 12 hours before use.

CASSIS & LEMON JAM
8
9
10
11
CHESTNUT TOPPING CREAM
12
13
14

카시스 & 레몬 잼

8. 설탕과 NH펙틴을 섞어준다.
9. 냄비에 카시스 퓌레와 레몬주스를 넣고 45℃로 데운다.
10. **9**에 **8**을 넣고 끓어오를 때까지 가열한 후 불에서 내린다.
11. 완성된 카시스&레몬 잼은 차갑게 식혀 사용한다.

체스트넛 토핑 크림

12. 밤 페이스트와 밤 퓌레를 섞어준다.
13. 기호에 따라 럼 또는 시럽을 선택해 조금씩 추가하며 섞어준다.
14. 파이핑하기 좋은 상태가 되면 마무리한다.

CASSIS & LEMON JAM

8. Mix sugar and pectin NH.
9. In a saucepan, heat cassis purée and lemon juice to 45°C.
10. Whisk in sugar mixture(**8**) to cassis mixture(**9**) until it starts to boil. Remove from heat.
11. Cool finished cassis & lemon jam completely before use.

CHESTNUT TOPPING CREAM

12. Mix chestnut paste and chestnut purée.
13. Rum or syrup can be added according to preference.
14. Continue to add until the texture is soft enough to pipe.

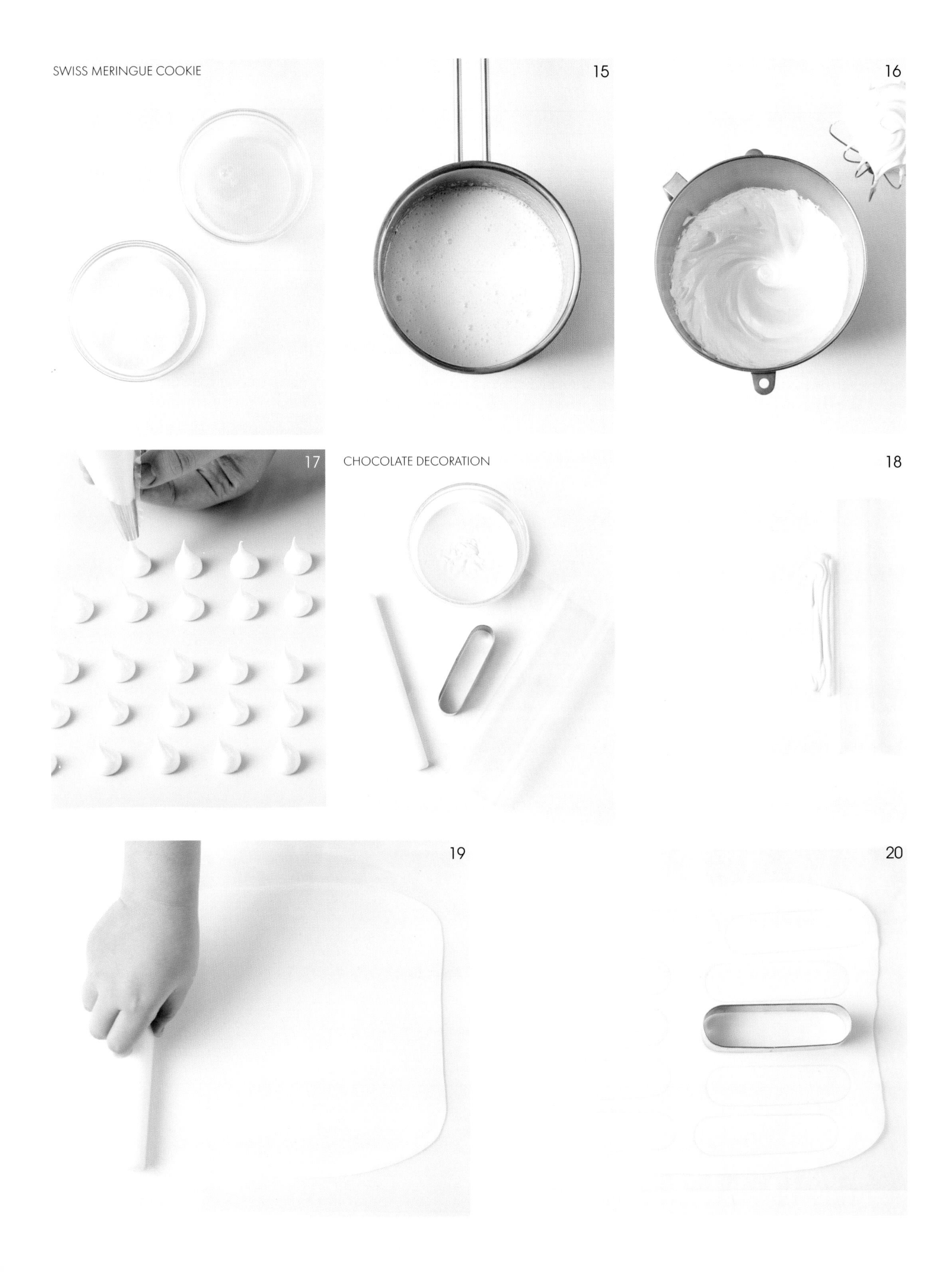
SWISS MERINGUE COOKIE
15
16
17
CHOCOLATE DECORATION
18
19
20

스위스 머랭 쿠키

15. 냄비에 설탕과 달걀흰자를 넣고 약불로 60~65℃가 될 때까지 가열한다.
16. 믹싱볼에 옮겨 고속으로 휘핑하다가 머랭이 하얗게 올라오고 힘 있는 상태가 되면 중속으로 낮춰 30℃가 될 때까지 식혀준다.
17. 원형(지름 1cm) 깍지를 이용해 원뿔 모양으로 파이핑한 후 70℃로 예열된 오븐에서 약 1시간 동안 건조시킨다.

초콜릿 장식물

18. 두 장의 투명 필름 사이에 템퍼링한 화이트초콜릿을 적당량 부어준다.
19. 필름을 덮고 밀대를 이용해 균일한 두께가 되도록 밀어편다.
20. 길이 12.5cm, 폭 3cm 에클레어 모양 커터로 커팅한다.

SWISS MERINGUE COOKIE

15. In a saucepan, heat sugar and egg whites until the temperature reaches 60~65°C.
16. Pour into a mixing bowl, and whip on high speed until stiff peak forms. Reduce to medium speed and continue to whip until the mixture cools down to 30°C.
17. Pipe the meringue into cones using a 1cm round tip. Dry in an oven preheated to 70°C, for about 1 hour.

CHOCOLATE DECORATION

18. Pour tempered white chocolate between two sheets of plastic film.
19. Roll out to even thickness using a rolling pin.
20. Cut into an éclair shape of 12.5cm long and 3cm wide.

FINISH

21

22

23

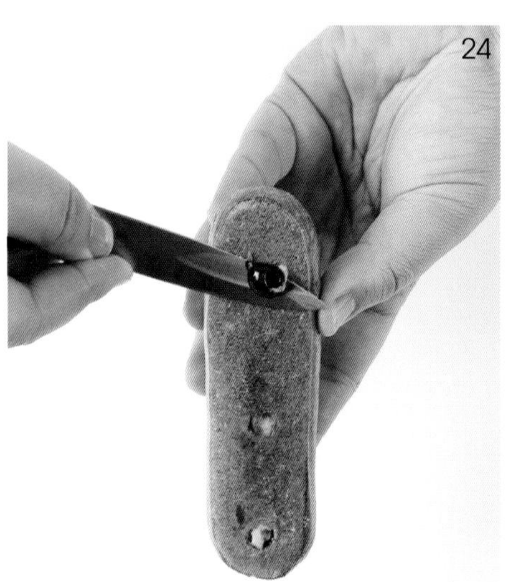
24

25

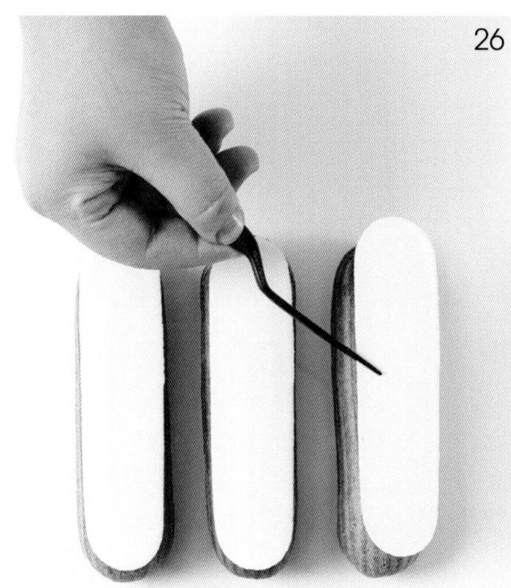
26

27

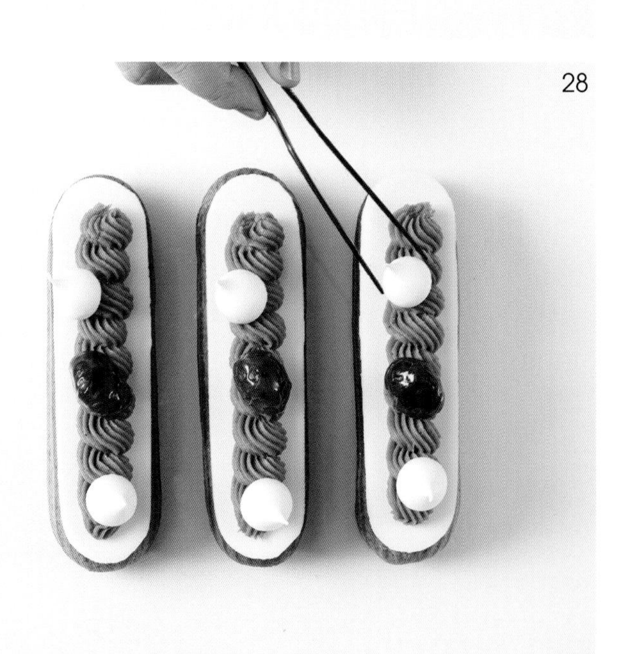
28

마무리

21. 구워낸 에클레어 바닥 부분에 작은 원형 깍지를 이용해 3개의 크림 주입구를 만든다.
22. 체스트넛 크림을 90% 정도 채운다.
23. 남은 공간에 카시스&레몬 잼을 가득 채운다.
24. 주입구를 깔끔하게 정리한다.
25. 에클레어 윗면에 여분의 체스트넛 크림을 파이핑한다.
26. 초콜릿 장식물을 올려 고정시킨다.
27. 톱니 모양(PF10번) 깍지를 이용해 체스트넛 토핑 크림을 파이핑한다.
28. 밤과 스위스 머랭 쿠키를 올려 마무리한다.

FINISH

21. Using a small round piping tip, make three holes on the bottom of the baked éclairs.
22. Fill with chestnut cream about 90% of the cavity.
23. Pipe in cassis & lemon jam in the remaining space.
24. Scrape off the excess around the holes.
25. Pipe some of the remaining chestnut cream on the top of the éclairs.
26. Fix the chocolate decoration piece over the cream.
27. Pipe chestnut cream topping using a star nozzle (PF10).
28. Place chestnuts and Swiss meringue cookies to finish.

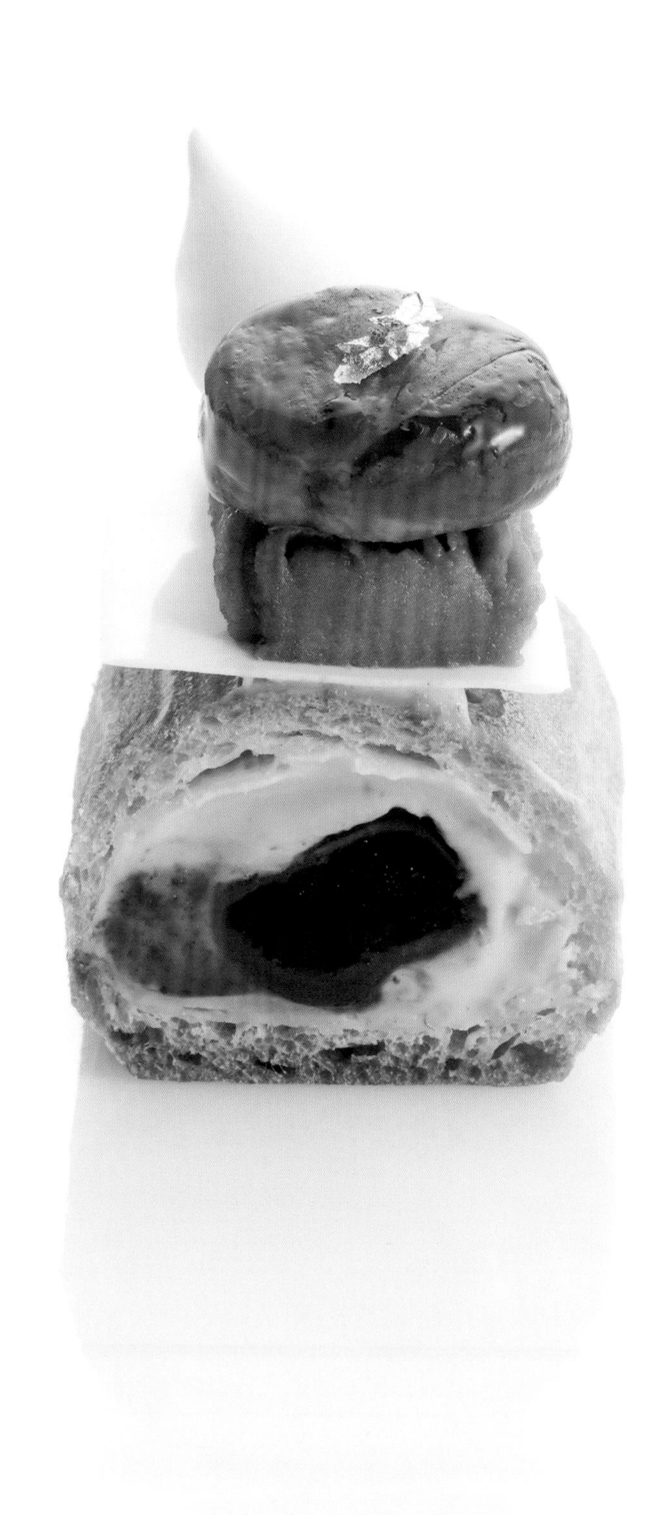

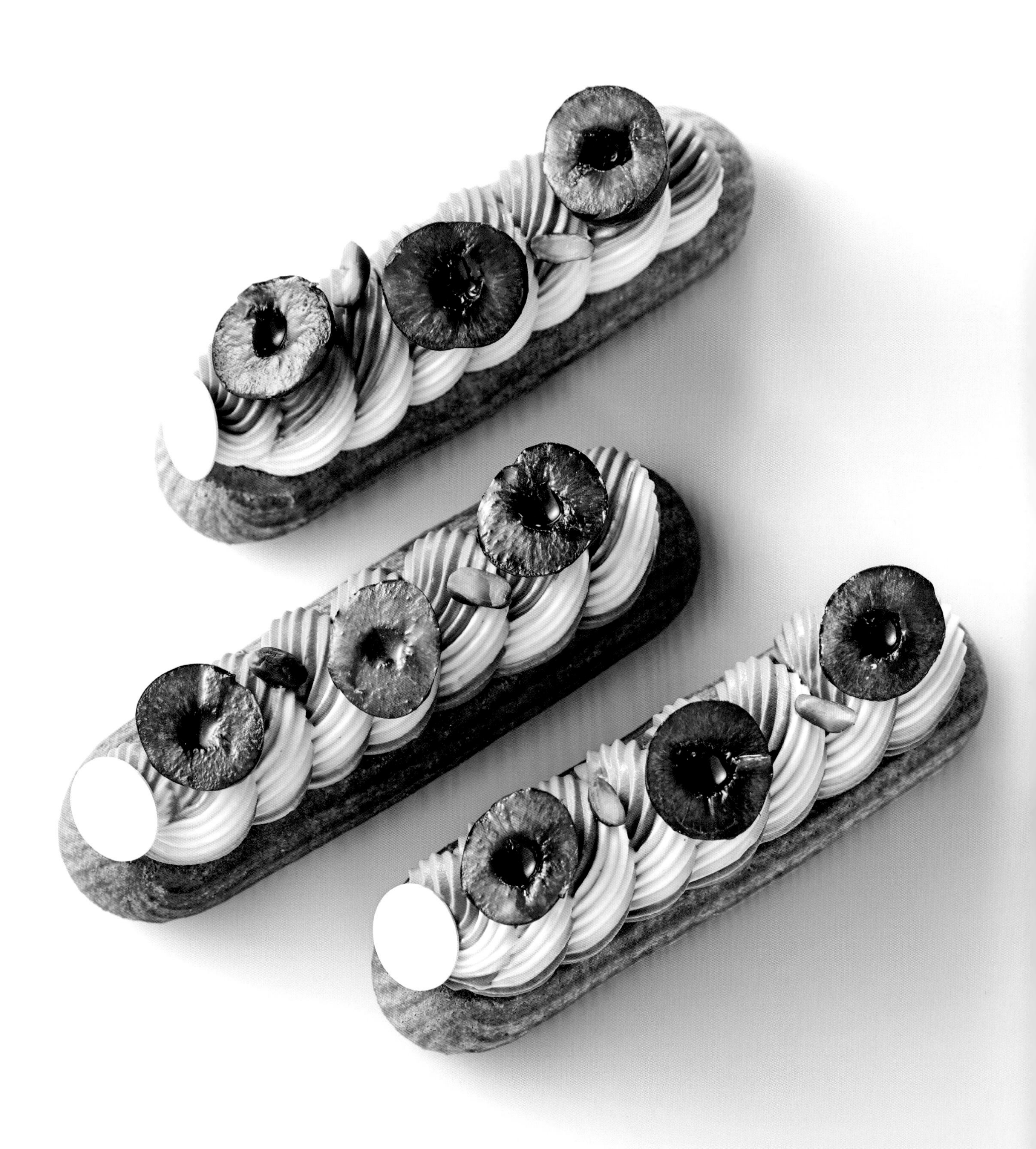

8 FORET NOIRE ECLAIR

포레누아 에클레어

ingredients - 15ea

파트 아 슈

BASIC GLUTEN FREE

비터 초콜릿 크림

(157p)

사워체리 젤

물 42g
사워체리 퓌레 101g
레몬주스 22g
아가아가 2g
설탕 33g

사워체리 인서트

사워체리 젤 163g
체리 237g

바닐라 휩 크림

우유 81g
바닐라빈 1/2개
젤라틴매스 12g
화이트초콜릿 72g
(IVOIRE 35%)
생크림 246g

다크초콜릿 휩 크림

우유 73g
젤라틴매스 12g
다크초콜릿 80g
(MANJARI 64%)
생크림 246g

장식물

체리
피스타치오

PATE A CHOUX

BASIC GLUTEN FREE

BITTER CHOCOLATE CREAM

(157p)

SOUR CHERRY GEL

42g Water
101g Sour cherry purée
22g Lemon juice
2g Agar-agar
33g Sugar

SOUR CHERRY INSERT

163g Sour cherry gel
237g Cherries

VANILLA WHIPPED CREAM

81g Milk
1/2pc Vanilla bean
12g Gelatin mass
72g White chocolate
(IVOIRE 35%)
246g Fresh cream

DARK CHOCOLATE WHIPPED CREAM

73g Milk
12g Gelatin mass
80g Dark chocolate
(MANJARI 64%)
246g Fresh cream

DECORATION

Cherries
Pistachios

SOUR CHERRY GEL
1
2
3
VANILLA WHIPPED CREAM
4
5
6
7

Process

사워체리 젤

1. 냄비에 물과 사워체리 퓌레를 넣고 45℃로 데운다.
2. 미리 섞어둔 설탕과 아가아가를 **1**에 넣고 끓어오를 때까지 가열한 후 볼에 옮겨 냉장고에서 굳힌다.
3. 핸드블렌더를 이용해 부드러운 젤 상태가 될 때까지 갈아준다.

바닐라 휩 크림

4. 끓기 직전까지 가열한 우유에 바닐라빈을 넣고 향을 우린 후 젤라틴매스를 혼합한다.
5. 35℃로 녹인 화이트초콜릿에 **4**를 체로 걸러 부어준 후 핸드블렌더로 혼합한다.
6. 35℃로 데운 생크림을 넣고 계속해서 핸드블렌더로 혼합한다.
7. 냉장고에서 12시간 휴지시킨 후 휘핑해 사용한다.

SOUR CHERRY GEL

1. Heat water and sour cherry purée in a saucepan to 45°C.
2. Whisk in previously mixed sugar and agar-agar into cherry purée mixture(**1**), and continue to boil. Remove the mixture into a bowl and set in the fridge.
3. Once it's set, blend with a hand blender to make a soft gel.

VANILLA WHIPPED CREAM

4. Heat milk and vanilla bean until just before boiling. Stir in gelatin mass.
5. Strain the milk mixture(**4**) over the white chocolate melted to 35°C. Mix with a hand blender.
6. Add fresh cream heated to 35°C, and continue to blend.
7. Rest in the fridge for 12 hours. Whip to use.

DARK CHOCOLATE WHIPPED CREAM

다크초콜릿 휩 크림

8. 끓기 직전까지 가열한 우유에 젤라틴매스를 혼합한 다음 35℃로 녹인 다크초콜릿에 붓고 핸드블렌더로 혼합한다.
9. 35℃로 데운 생크림을 넣고 계속해서 핸드블렌더를 이용해 혼합한다.
10. 냉장고에서 12시간 휴지시킨 후 휘핑해 사용한다.

사워체리 인서트

11. 사워체리 젤과 사방 3mm로 작게 자른 체리를 섞어 사워체리 인서트를 완성한다.

마무리

12. 구워낸 에클레어 윗면을 제거한다.
13. 비터 초콜릿 크림을 20g씩 채운다.
14. 사워체리 인서트를 가득 채운다.
15. 톱니 모양(PF12번) 깍지를 끼운 짤주머니에 바닐라 휩 크림, 다크초콜릿 휩 크림을 절반씩 담는다.
16. 에클레어 윗면에 파이핑한다.
17. 체리와 피스타치오로 장식해 마무리한다.

DARK CHOCOLATE WHIPPED CREAM

8. Heat milk until just before boiling, and stir in gelatin mass. Pour over dark chocolate melted to 35°C, and mix using a hand blender.
9. Add fresh cream heated to 35°C. Continue to blend.
10. Set in the fridge for 12 hours. Whip to use.

SOUR CHERRY INSERT

11. Mix sour cherry gel and cherries cut into 3mm cube to make sour cherry insert.

FINISH

12. Cut the top of the baked éclairs.
13. Fill 20g of bitter chocolate cream.
14. Fill the rest with the sour cherry insert.
15. In a piping bag with a star nozzle (PF12), fill half of the bag with vanilla whipped cream and the other half with dark chocolate whipped cream.
16. Pipe on the top of the éclairs.
17. Decorate with cherries and pistachios to finish.

9 PISTACHIO ECLAIR

피스타치오 에클레어

ingredients - 15ea

파트 아 슈

BASIC GLUTEN FREE

장식물

피스타치오
게랑드소금

피스타치오 페이스트*

피스타치오 252g
설탕 126g
물 42g
게랑드소금 2.5g

피스타치오 크림

달걀노른자 34g
설탕 58g
전분 22g
우유 413g
피스타치오 페이스트* 54g
젤라틴매스 16.8g
버터 116g

글레이즈

생크림 183g
물엿 73g
젤라틴매스 42g
화이트초콜릿 231g
(OPALYS 33%)
화이트 코팅초콜릿 207g
노란색 천연 식용 색소 적당량
녹색 천연 식용 색소 적당량

PATE A CHOUX

BASIC GLUTEN FREE

DECORATION

Pistachios
Guerande salt

PISTACHIO PASTE*

252g Pistachios
126g Sugar
42g Water
2.5g Guerande salt

PISTACHIO CREAM

34g Egg yolks
58g Sugar
22g Starch
413g Milk
54g Pistachio paste*
16.8g Gelatin mass
116g Butter

GLAZE

183g Fresh cream
73g Corn syrup
42g Gelatin mass
231g White chocolate
(OPALYS 33%)
207g White compound chocolate
Yellow food coloring (natural) QS
Green food coloring (natural) QS

1
2
3-1
3-2
4
5-1
5-2
6

Process

피스타치오 페이스트

1. 피스타치오는 170℃로 예열된 오븐에서 약 15분간 구워준다. 냄비에 물과 설탕을 넣고 118~121℃까지 끓인다.
2. 118~121℃가 되면 불에서 내린 다음 따뜻한 상태의 피스타치오를 넣고 섞어준다.
3. 시럽이 하얗게 재결정화 상태가 될 때까지 계속해서 저어준다.
4. 실팻에 넓게 펼친 후 완전히 식힌다.
5. 푸드프로세서에 식힌 피스타치오와 게랑드소금을 넣고 갈아준다.
6. 흐르는 정도의 페이스트 상태가 되면 마무리한다.

PISTACHIO PASTE

1. Preheat oven to 170°C. Roast pistachios for about 15 minutes. In a saucepan, boil water and sugar until the temperature reaches 118~121°C.
2. Once it reaches 118~121°C, remove from heat and mix with the warm pistachios.
3. Continue to stir until the syrup recrystallizes and turns white.
4. Spread over Silpat to cool completely.
5. In a food processor, grind cooled pistachios and Guerande salt.
6. Finish when the mixture turns into a fluid paste.

7
8
9
10
11
12
13
14

피스타치오 크림

7. 볼에 달걀노른자, 설탕 1/2, 전분, 피스타치오 페이스트를 넣고 섞는다.
8. 냄비에 우유와 나머지 설탕을 넣고 45℃로 데운 후 **7**에 조금씩 부어가며 섞는다.
9. 체에 걸러 다시 냄비로 옮겨준다.
10. **9**가 충분히 호화되도록 가열한다.
11. 볼에 옮긴 후 젤라틴매스를 혼합한다.
12. 45℃까지 식힌다.
13. 상온 상태의 버터를 넣고 핸드블렌더로 혼합한다.
14. 냉장고에서 12시간 휴지시켜 사용한다.

PISTACHIO CREAM

7. In a bowl, mix egg yolks, half of the sugar, starch, and pistachio paste.
8. Heat milk and remaining sugar to 45°C. Gradually add into the pistachio mixture(**7**) and mix to incorporate.
9. Strain the mixture and pour it back into the saucepan.
10. Heat until gelatinized enough.
11. Transfer to a bowl and stir in gelatin mass.
12. Cool down to 45°C.
13. Mix with room temperature butter using a hand blender.
14. Rest in the fridge for 12 hours before use.

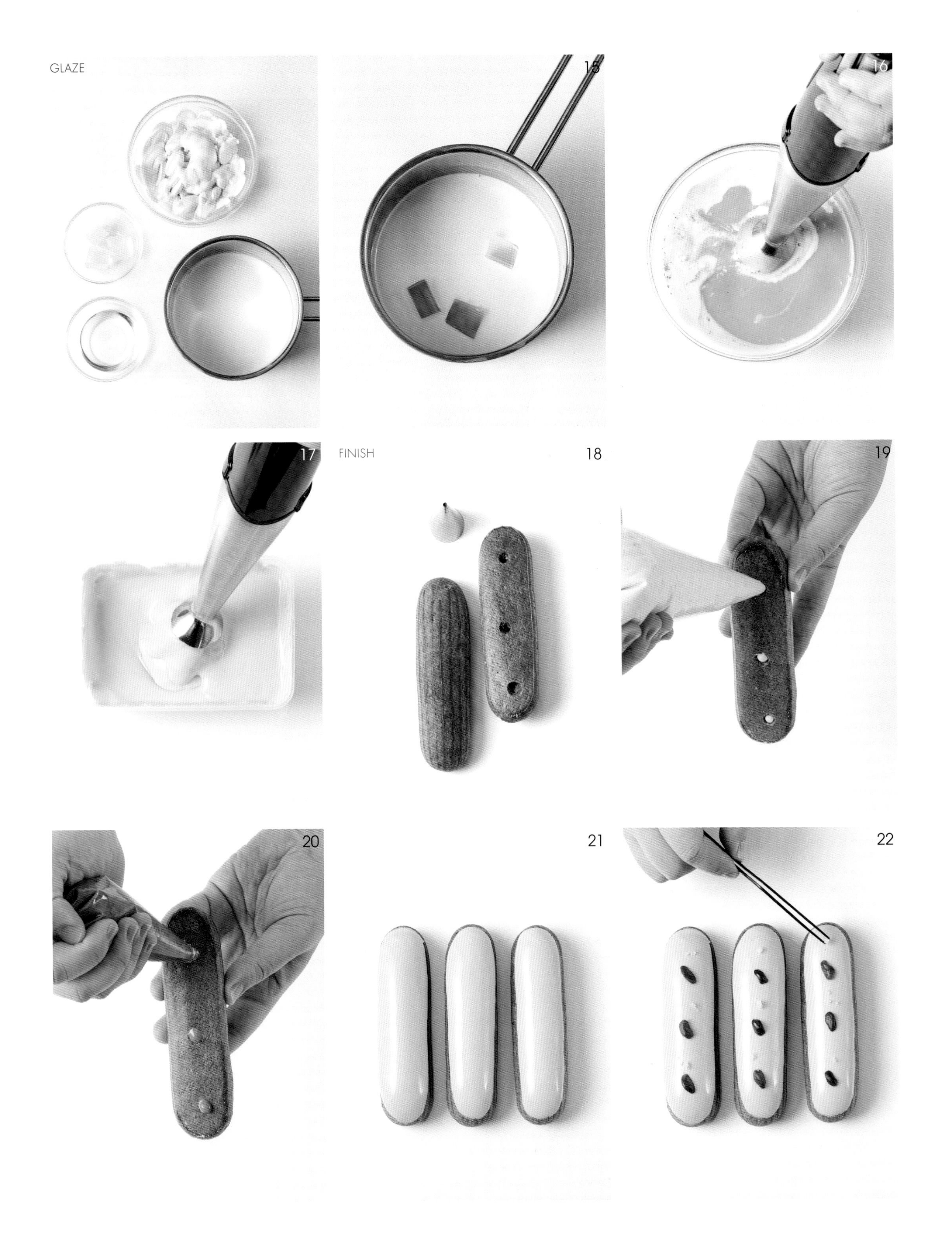
GLAZE
15
16
17
FINISH
18
19
20
21
22

글레이즈

15. 생크림과 물엿을 끓기 직전까지 가열한 후 젤라틴매스를 혼합한다.
16. 35℃로 녹인 화이트초콜릿과 코팅초콜릿에 15를 붓고 핸드블렌더로 혼합한다. 이때 노란색과 녹색 천연 식용 색소를 첨가한다. 냉장고에서 12시간 세팅시킨다.
17. 사용할 때는 다시 30℃로 온도를 맞춘 다음 핸드블렌더로 균일하게 믹싱해 사용한다.

마무리

18. 구워낸 에클레어 바닥 부분에 작은 원형 깍지를 이용해 3개의 크림 주입구를 만든다.
19. 피스타치오 크림을 90% 정도 채운다.
20. 남은 공간에 피스타치오 페이스트를 가득 채운 후 주입구를 깔끔하게 정리한다.
21. 에클레어 윗면을 글레이징한다.
22. 피스타치오와 게랑드소금을 올려 마무리한다.

GLAZE

15. Heat fresh cream and corn syrup until just before boiling, then stir in gelatin mass.
16. Mix the cream mixture(**15**) to white chocolate melted to 35°C using a hand blender. Add yellow and green natural food coloring. Set in the fridge for 12 hours.
17. To use, warm the glaze to 30°C and mix with a hand blender thoroughly.

FINISH

18. Using a small round piping tip, make three holes on the bottom of the baked éclairs.
19. Fill with pistachio cream about 90% of the cavity.
20. Fill with pistachio paste in the remaining space. Scrape off the excess around the holes.
21. Glaze the top of the éclairs.
22. Decorate with pistachios and Guerande salt to finish.

10 # BLACK SESAME ECLAIR

흑임자 에클레어

ingredients - 15ea

파트 아 슈

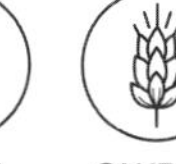

BASIC GLUTEN FREE

장식물

깨 튀일*

흑임자 크림

달걀노른자 34g
설탕 58g
전분 22g
흑임자 페이스트 56g
우유 411g
오렌지제스트 2g
젤라틴매스 16.8g
버터 116g

글레이즈

생크림 183g
물엿 73g
젤라틴매스 42g
화이트초콜릿 231g
(OPALYS 33%)
화이트 코팅초콜릿 207g

깨 튀일*

참깨 70g
검정깨 70g
볶은 현미 20g
박력분 30g
달걀흰자 80g
설탕 56g
버터 40g
화이트 코팅초콜릿
카카오버터

PATE A CHOUX

BASIC GLUTEN FREE

DECORATION

Sesame Tuiles*

BLACK SESAME CREAM

34g Egg yolks
58g Sugar
22g Starch
56g Black sesame paste
411g Milk
2g Orange zest
16.8g Gelatin mass
116g Butter

GLAZE

183g Fresh cream
73g Corn syrup
42g Gelatin mass
231g White chocolate
(OPALYS 33%)
207g White compound chocolate

SESAMI TUILE*

70g Sesame seeds
70g Black sesame seeds
20g Brown rice, toasted
30g Cake flour
80g Egg white
56g Sugar
40g Butter
White compound chocolate
Cacao butter

BLACK SESAME CREAM
1
2
3
4
5
GLAZE
6
7

Process

흑임자 크림

1. 볼에 달걀노른자, 설탕 1/2, 전분, 흑임자 페이스트를 넣고 섞어준다.
2. 냄비에 우유와 나머지 설탕을 넣고 끓기 직전까지 가열한 후 오렌지제스트를 넣어 향을 우린다. 1에 조금씩 부어가며 혼합한다.
3. 체에 걸러 냄비에 옮긴 후 충분히 호화되도록 가열한다.
4. 젤라틴매스를 혼합한 다음 45℃까지 식혀준다.
5. 상온 상태의 버터를 넣고 핸드블렌더로 혼합한다. 냉장고에서 12시간 휴지시켜 사용한다.

글레이즈

6. 끓기 직전까지 가열한 생크림과 물엿에 젤라틴매스를 섞어준다. 35℃로 녹인 화이트초콜릿과 코팅초콜릿에 붓고 핸드블렌더로 혼합한 후 냉장고에서 12시간 세팅시킨다.
7. 사용할 때는 다시 30℃로 온도를 맞춘 다음 핸드블렌더로 균일하게 믹싱해 사용한다.

BLACK SESAME CREAM

1. In a bowl, mix egg yolks, half of the sugar, starch, and black sesame paste.
2. Heat milk and remaining sugar in a saucepan until just before boiling. Stir in the orange zest to infuse. Gradually pour over the egg mixture(1). Stir to incorporate.
3. Strain over a saucepan. Heat to cook.
4. Stir in gelatin mass and let it cool to 45°C.
5. Mix in room temperature butter with a hand blender. Set in the fridge for 12 hours before use.

GLAZE

6. Heat fresh cream and corn syrup until just before boiling. Stir in gelatin mass. Pour over white chocolate melted to 35°C and compound chocolate, and mix to incorporate using a hand blender. Set in the fridge for 12 hours.
7. To use, reheat the glaze to 30°C, and mix thoroughly using a hand blender.

SESAMI TUILE
8
9
10
11
12
FINISH
13
14
15

깨 튀일

8. 달걀흰자, 설탕, 체 친 박력분을 섞어준다.
9. 8에 참깨, 검정깨, 볶은 현미를 넣어 섞고 55~60℃로 녹인 버터를 혼합한다.
10. 스패출러를 이용해 실팻에 고르게 펼친 후, 180℃로 예열된 오븐에서 10분간 굽는다.
11. 오븐에서 나온 직후 튀일 표면에 녹인 카카오버터를 얇게 바른다.
12. 완전히 식힌 후 튀일 뒷면에 녹인 화이트 코팅초콜릿을 얇게 바른다.

마무리

13. 구워낸 에클레어 바닥 부분에 작은 원형 깍지를 이용해 3개의 크림 주입구를 만든 후 흑임자 크림을 가득 채운다.
14. 주입구를 깔끔하게 정리한 후 에클레어 윗면을 글레이징한다.
15. 깨 튀일을 올려 마무리한다.

SESAMI TUILE

8. Mix together egg whites, sugar, and sifted cake flour.
9. Add sesame seeds, black sesame seeds, and toasted brown rice to egg white mixture(**8**). Stir in butter melted to 55~60°C.
10. Spread evenly over Silpat using a spatula. Put in the preheated oven at 180°C, for about 10 minutes.
11. As soon as the tuiles are out of the oven, brush with melted cacao butter.
12. Cool completely. Apply a thin layer of melted white compound chocolate on the back of the tuiles.

FINISH

13. Using a small round piping tip, make three holes on the bottom of the baked éclairs. Fill with black sesame cream.
14. Scrap off the excess around the holes, and glaze the top of the éclairs.
15. Top with sesame tuiles to finish.

BITTER CHOCOLATE

11 BITTER CHOCOLATE ECLAIR

비터 초콜릿 에클레어

ingredients - 15ea

파트 아 슈

BASIC

GLUTEN FREE

비터 초콜릿 크림

달걀노른자 132g
설탕 67g
우유 181g
생크림 181g
다크초콜릿 141g
(CARAIBE 66%)

글레이즈

생크림 223g
물엿 67g
젤라틴매스 42g
다크초콜릿 213g
(CARAIBE 66%)
다크 코팅초콜릿 190g

장식물

다크초콜릿
(EQUATORIALE NOIRE 55%)
카카오버터
검정색 천연 식용 색소
식용금박

PATE A CHOUX

BASIC

GLUTEN FREE

BITTER CHOCOLATE CREAM

132g Egg yolks
67g Sugar
181g Milk
181g Fresh cream
141g Dark chocolate
(CARAIBE 66%)

GLAZE

223g Fresh cream
67g Corn syrup
42g Gelatin mass
213g Dark chocolate
(CARAIBE 66%)
190g Dark compound chocolate

DECORATION

Dark chocolate
(EQUATORIALE NOIRE 55%)
Cacao butter
Black food coloring (natural)
Edible gold leaf

BITTER CHOCOLATE CREAM
1
2
3
4
5
BRAUN
GLAZE
6
7

Process

비터 초콜릿 크림

1. 볼에 달걀노른자, 설탕 1/2을 넣고 혼합한다. 냄비에 생크림, 우유, 나머지 설탕을 넣고 45℃로 데운 후 달걀노른자 믹스처에 조금씩 부어가며 혼합한다.
2. 냄비에 옮겨 83~85℃로 가열해 크렘 앙글레이즈를 만든다.
3. 체에 걸러준 후 45℃까지 식힌다.
4. 35℃로 녹인 다크초콜릿에 **3**을 3~4회 나누어 섞는다.
5. 핸드블렌더로 혼합한 후 냉장고에서 12시간 휴지시켜 사용한다.

글레이즈

6. 끓기 직전까지 가열한 생크림과 물엿에 젤라틴매스를 넣고 섞어준다. 35℃로 녹인 다크초콜릿과 코팅초콜릿에 붓고 핸드블렌더로 혼합한 후 냉장고에서 12시간 세팅시킨다.
7. 사용할 때는 다시 30℃로 온도를 맞춘 다음 핸드블렌더로 균일하게 믹싱해 사용한다.

BITTER CHOCOLATE CREAM

1. In a bowl, mix egg yolks and half of the sugar. In a saucepan, heat fresh cream, milk, and remaining sugar to 45°C. Gradually add into the egg mixture to incorporate.
2. Pour back into a saucepan and boil until the temperature reaches 83~85°C to make crème anglais.
3. Strain and cool down to 45°C.
4. Mix dark chocolate melted to 35°C to crème anglais(**3**) in 3~4 times.
5. Incorporate with a hand blender, and set in the fridge for 12 hours before use.

GLAZE

6. Heat fresh cream and corn syrup until just before boiling, and stir in gelatin mass. Pour over dark chocolate melted to 35°C and compound chocolate. Mix with a hand blender and set in the fridge for 12 hours.
7. To use, reheat the glaze to 30°C, and mix thoroughly using a hand blender.

8
9
10
11
12
BITTER CHOCOLATE
BITTER CHOCOLATE
BITTER CHOCOLATE

FINISH

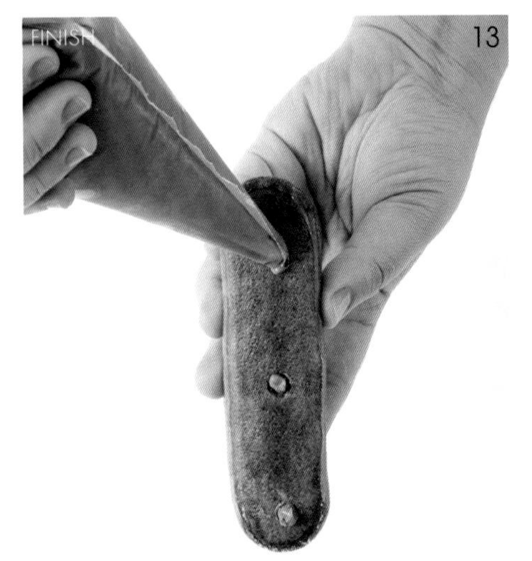
13

14

15
BITTER CHOCOLATE

초콜릿 장식물

8. 디자인한 필름 위에 템퍼링한 다크초콜릿을 적당량 부어준다.
9. 스패출러를 이용해 다크초콜릿을 1mm 두께로 얇게 펴준다. 이때 실리콘 재질의 두께 바를 이용하면 일정한 두께로 밀어펼 수 있다.
10. 초콜릿이 굳기 시작하면 조심스럽게 필름을 들어올린다.
11. 충분히 수축시킨 후 필름을 떼어내고 냉동고에 30분 동안 넣어둔다.
12. 다크초콜릿과 카카오버터를 1:1 비율로 혼합한 후 차가워진 초콜릿 표면에 분사해 벨벳 같은 느낌을 표현한다.

마무리

13. 구워낸 에클레어 바닥 부분에 작은 원형 깍지를 이용해 3개의 크림 주입구를 만든 후 비터 초콜릿 크림을 가득 채운다.
14. 주입구를 깔끔하게 정리한 후 에클레어 윗면을 글레이징한다.
15. 초콜릿 장식물을 올려 마무리한다.

CHOCOLATE DECORATION

8. Pour a small amount of tempered dark chocolate over the stencil.
9. Spread the chocolate thinly to 1mm using a spatula. using a thin silicon bars will help to spread evenly.
10. When chocolate starts to crystallize, carefully lift the stencil.
11. Give enough time to crystallize completely before removing the stencil. Store the chocolate in the freezer for 30 minutes.
12. Mix 1-part dark chocolate with 1-part cacao butter, and spray over the cold surface of chocolate piece to give a velvety texture.

FINISH

13. Using a small round piping tip, make three holes on the bottom of the baked éclairs. Fill with bitter chocolate cream.
14. Scrap off the excess around the holes, and glaze the top of the éclairs.
15. Top with the sprayed chocolate piece to finish.

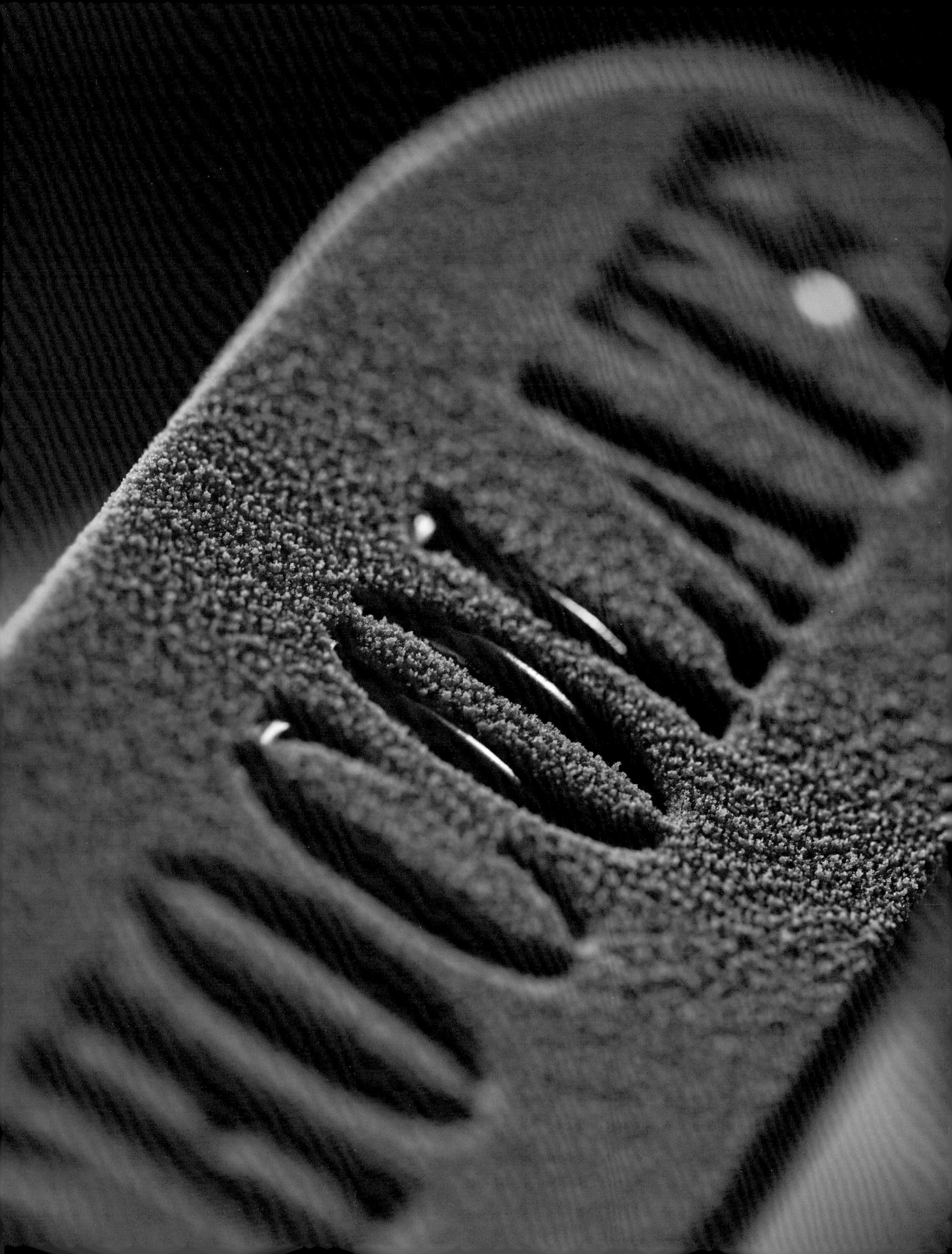

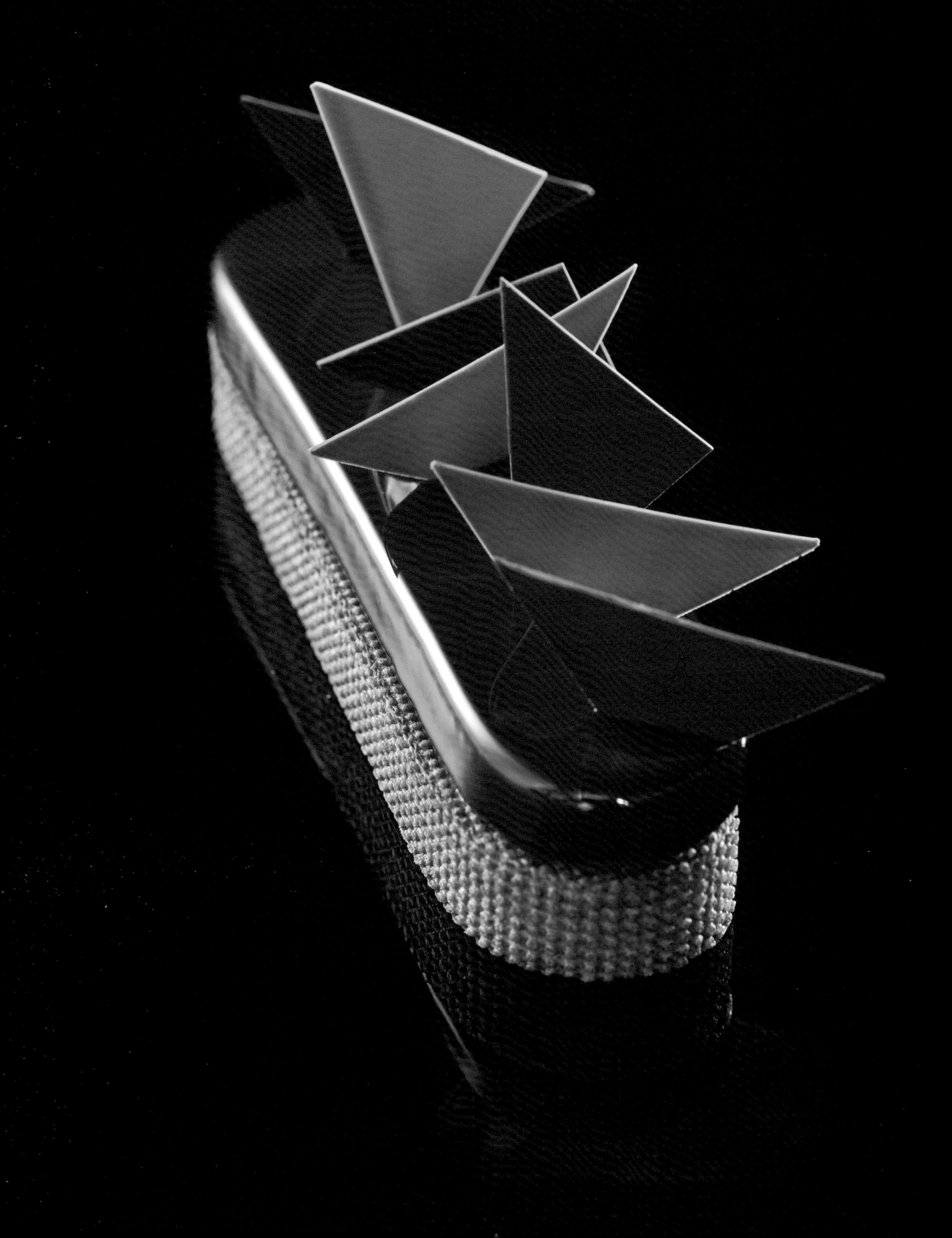

12 HAZELNUT CHOCOLATE BAR

헤이즐넛 초콜릿 바

ingredients - 15ea

파트 아 슈

BASIC GLUTEN FREE

장식물

다크초콜릿
(EQUATORIALE NOIRE 55%)

밀크초콜릿
(JIVARA LATTE 40%)

헤이즐넛 카카오 프랄리네

헤이즐넛 250g
카카오닙 25g
설탕 250g
물 58g
게랑드소금 2.5g

다크초콜릿 무스

우유 76g
설탕 21g
젤라틴매스 12g
다크초콜릿 108g
(CARAIBE 66%)
생크림 194g

글레이즈

물 75g
설탕 150g
물엿 150g
연유 100g
젤라틴매스 60g
카카오파우더 30g
(POUDRE DE CACAO)
다크초콜릿 120g
(EQUATORIALE NOIRE 55%)

PATE A CHOUX

BASIC GLUTEN FREE

DECORATION

Dark chocolate
(EQUATORIALE NOIRE 55%)

Milk chocolate
(JIVARA LATTE 40%)

HAZELNUT CACAO PRALINE

250g Hazelnuts
25g Cacao nips
250g Sugar
58g Water
2.5g Guerande salt

DARK CHOCOLATE MOUSSE

76g Milk
21g Sugar
12g Gelatin mass
108g Dark chocolate
(CARAIBE 66%)
194g Fresh cream

GLAZE

75g Water
150g Sugar
150g Corn syrup
100g Condensed milk
60g Gelatin mass
30g Cacao powder
(POUDRE DE CACAO)
120g Dark chocolate
(EQUATORIALE NOIRE 55%)

1
2
3
4
5
6
7
8

Process

헤이즐넛 카카오 프랄리네

1. 헤이즐넛은 140℃로 예열된 오븐에서 약 8분간 구워준다. 냄비에 물과 설탕을 넣고 118~121℃까지 끓인다.
2. 118~121℃가 되면 불에서 내린 후 따뜻한 상태의 헤이즐넛을 넣고 섞어준다.
3. 시럽이 하얗게 재결정화 상태가 될 때까지 계속해서 저어준다.
4. 다시 불에 올려 골고루 저어주며 계속해서 가열한다.
5. 캐러멜화가 시작되면 카카오닙을 넣고 진한 갈색이 날 때까지 조금 더 진행한다.
6. 실팻에 넓게 펼쳐 완전히 식힌다.
7. 푸드프로세서에 식힌 캐러멜라이즈 헤이즐넛 카카오와 게랑드소금을 넣고 갈아준다.
8. 흐르는 정도의 페이스트 상태가 되면 마무리한다.

HAZELNUT CACAO PRALINE

1. Preheat oven to 140°C. Roast hazelnuts for about 8 minutes. Boil water and sugar in a saucepan until it reached 118~121°C.
2. When the temperature reaches 118~121°C, remove from heat and mix with warm hazelnuts.
3. Continue to stir until the syrup recrystallizes and turns white.
4. Place it back over the heat, and continue to stir.
5. When it starts to caramelize, mix in cacao nibs, and continue to cook until the color darkens a bit more.
6. Spread over Silpat, and cool completely.
7. Grind cooled caramelized hazelnut cacao and Guerande salt in a food processor.
8. Finish when the texture turns into a fluid paste.

9
10
11-1
11-2
12
13
14
15

다크초콜릿 무스

9. 생크림은 무스 상태로 휘핑한 후 16~18℃로 준비한다.
10. 냄비에 우유와 설탕을 넣고 끓기 직전까지 가열한 후 젤라틴매스를 혼합한다.
11. 35℃로 녹인 다크초콜릿에 **10**을 붓고 핸드블렌더로 혼합해 가나슈 베이스를 만든다.
12. 휘핑한 생크림과 45℃의 가나슈 베이스를 혼합한다.
13. 완성된 무스는 짤주머니에 담는다.
14. 타원형 몰드를 실팻 위에 올려 준비한다.
15. 몰드 안에 완성한 다크 초콜릿 무스를 채운 후 급속 냉동고에서 단단하게 굳힌다.

DARK CHOCOLATE MOUSSE

9. Whip cream to mousse consistancy, and keep at 16~18°C.
10. Heat milk and sugar until just before boiling, and stir in gelatin mass.
11. Pour milk mixture(**10**) into dark chocolate melted to 35°C, and incorporate using a hand blender to make a base for the ganache.
12. When the ganache base cools to 45°C, fold in the whipped cream.
13. Fill a piping bag with the mousse.
14. Prepare oblong mold over Silpat.
15. Fill the mold with dark chocolate mousse and blast freeze until completely frozen.

GLAZE
16
17
18-1
18-2
19-1
19-2
20

글레이즈

16. 냄비에 물과 설탕을 넣고 가열하다가 시럽 상태가 되면 물엿을 넣고 103℃까지 끓인다.
17. 카카오파우더를 넣고 섞어준다.
18. 연유와 젤라틴매스를 순서대로 넣고 혼합한다.
19. 35℃로 녹인 다크초콜릿에 붓고 핸드블렌더로 혼합한 후 냉장고에서 12시간 세팅시킨다.
20. 사용할 때는 다시 35℃로 온도를 맞춘 다음 핸드블렌더로 균일하게 믹싱해 사용한다.

GLAZE

16. Boil water and sugar to make syrup. Add corn syrup and continue to boil until it reaches 103°C.
17. Mix in cacao powder.
18. Add condensed milk and gelatin mass in order, and incorporate.
19. Pour over dark chocolate melted to 35°C, and mix using a hand blender. Set in the fridge for 12 hours.
20. To use, reheat the glaze to 30°C, and mix thoroughly using a hand blender.

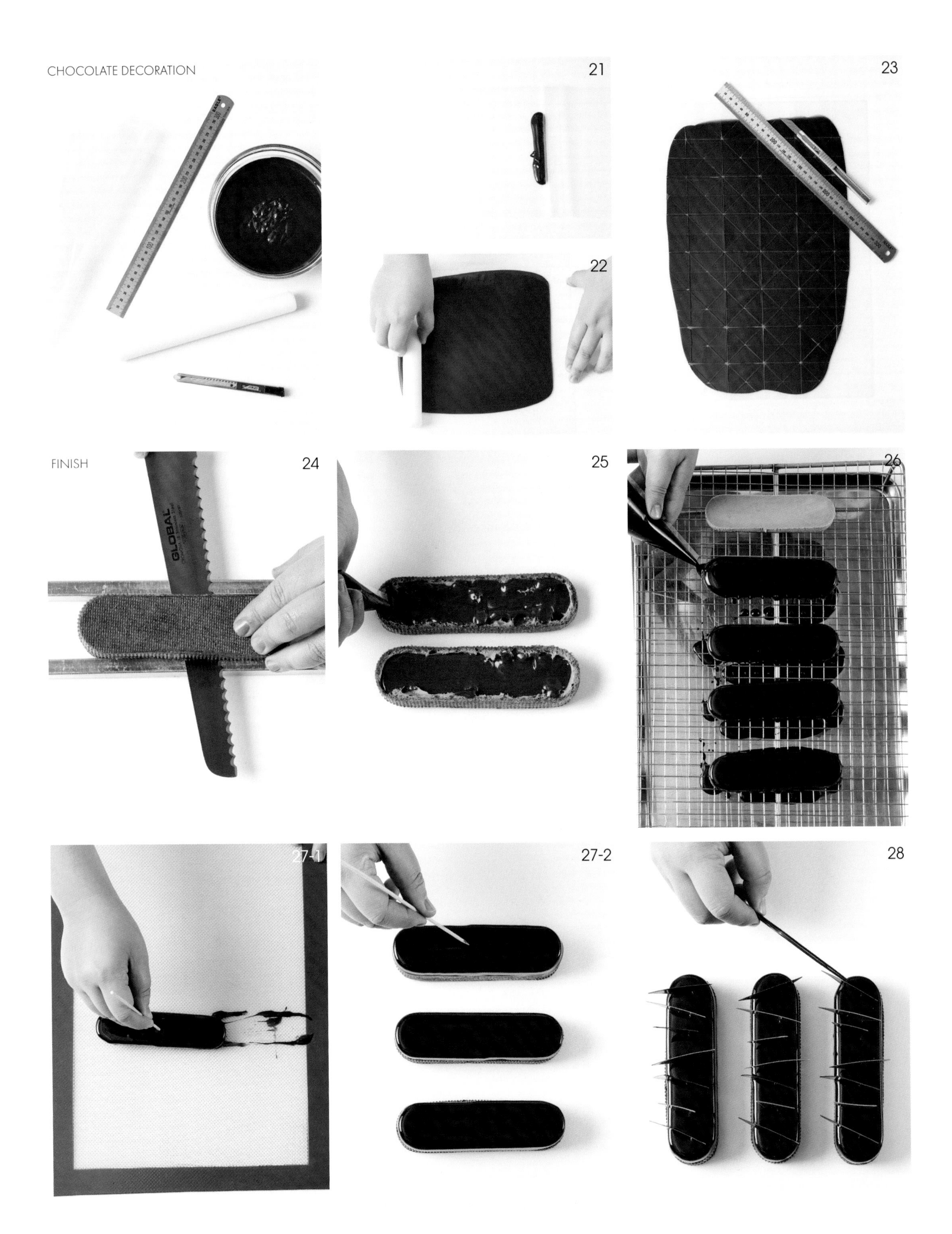
CHOCOLATE DECORATION
21
22
23
FINISH
24
GLOBAL
25
26
27-1
27-2
28

초콜릿 장식물

21. 두 장의 투명 필름 사이에 템퍼링한 다크초콜릿 적당량을 부어준다.
22. 필름을 덮고 밀대를 이용해 균일한 두께가 되도록 밀어편다.
23. 2cm 정삼각형 모양으로 커팅한다.

마무리

24. 블럭 형태로 구워낸 에클레어는 충분히 식힌 후 높이 1.5cm가 되도록 반으로 자른다. (47p)
25. 헤이즐넛 카카오 프랄린을 가득 채운다.
26. 단단하게 굳은 다크초콜릿 무스를 글레이징한다.
27. 흐르는 부분을 깔끔하게 정리한 후 에클레어 위에 얹어준다.
28. 초콜릿 장식물을 얹어 완성한다.

CHOCOLATE DECORATION

21. Place a small amount of tempered dark chocolate between two sheets of plastic film.
22. Roll into even thickness using a rolling pin.
23. Cut into 2cm triangles.

FINISH

24. Bake éclairs into a block shape. Cool completely. Cut in half to make a height of 1.5cm. (47p)
25. Fill with hazelnut cacao praline.
26. Glaze the frozen dark chocolate mousse.
27. Clean the bottom of the glazed mousse and place it on top of the éclairs.
28. Decorate with chocolate decorations to finish.

Aroma

EARL GREY ECLAIR

BROWN RICE GREEN TEA ECLAIR

ROSE ECLAIR

COFFEE ECLAIR

SALTED BUTTER CARAMEL ECLAIR

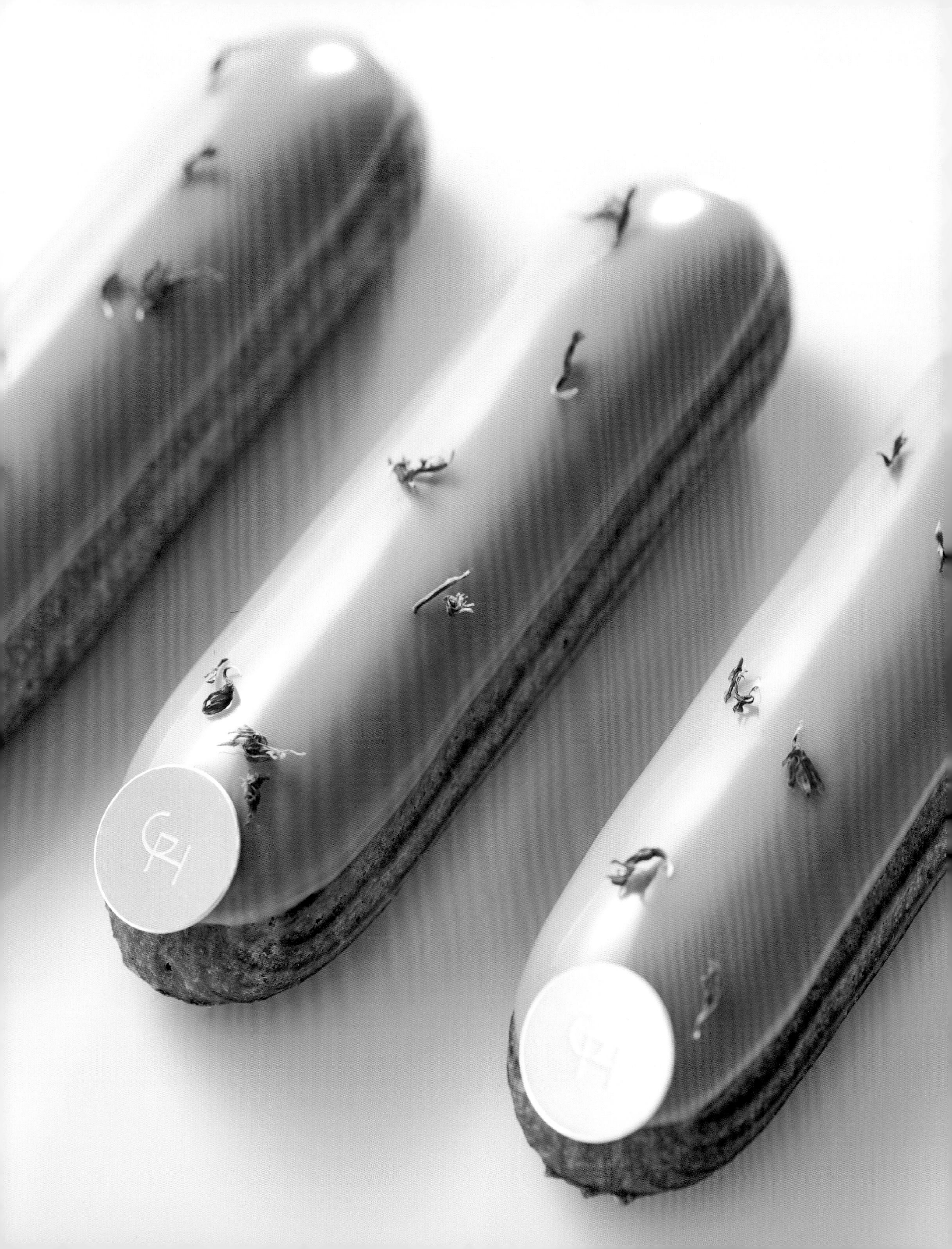

13 EARL GREY ECLAIR

얼그레이 에클레어

ingredients - 15ea

파트 아 슈

BASIC

GLUTEN FREE

얼그레이 크림

달걀노른자 62g
설탕 46g
전분 18g
우유 314g
생크림 69g
얼그레이 잎 7g
블론드초콜릿 123g
(DULCEY 32%)
버터 69g

글레이즈

생크림 183g
물엿 73g
젤라틴매스 42g
블론드초콜릿 231g
(DULCEY 32%)
화이트 코팅초콜릿 207g

장식물

콘플라워

PATE A CHOUX

BASIC

GLUTEN FREE

EARL GREY CREAM

62g Egg yolks
46g Sugar
18g Starch
314g Milk
69g Fresh cream
7g Earl Grey tea leaves
123g Blonde chocolate
(DULCEY 32%)
69g Butter

GLAZE

183g Fresh cream
73g Corn syrup
42g Gelatin mass
231g Blonde chocolate
(DULCEY 32%)
207g White compound chocolate

DECORATION

Cornflowers

EARL GREY CREAM
1
2
3
4
5
6
7
8

Process

얼그레이 크림

1. 달걀노른자, 설탕 1/2, 전분을 혼합한다.
2. 냄비에 우유, 생크림, 나머지 설탕을 넣고 끓기 직전까지 가열한 후 얼그레이 잎을 넣어 향을 우린다.
3. **1**에 **2**를 조금씩 부어가며 섞어준다.
4. 체에 걸러 다시 냄비로 옮긴다.
5. 충분히 호화될 때까지 가열한다.
6. 45℃까지 식힌다.
7. 35℃로 녹인 블론드초콜릿에 붓고 핸드블렌더로 혼합한다.
8. 상온 상태의 버터를 넣고 계속해서 핸드블렌더로 혼합한 후 냉장고에서 12시간 휴지시켜 사용한다.

EARL GREY CREAM

1. Mix egg yolks, half of the sugar, and starch.
2. In a saucepan, heat milk, fresh cream, and remaining sugar until just before boiling. Add Earl Grey tea leaves to infuse.
3. Gradually stir in the infused mixture(**2**) into egg yolks mixture(**1**), a little bit at a time.
4. Strain into a saucepan.
5. Heat until gelatinized enough.
6. Cool to 45°C.
7. Pour onto Blonde chocolate melted to 35°C, and incorporate using a hand blender.
8. Add room temperature butter and continue to mix with a hand blender. Rest in the fridge for 12 hours before use.

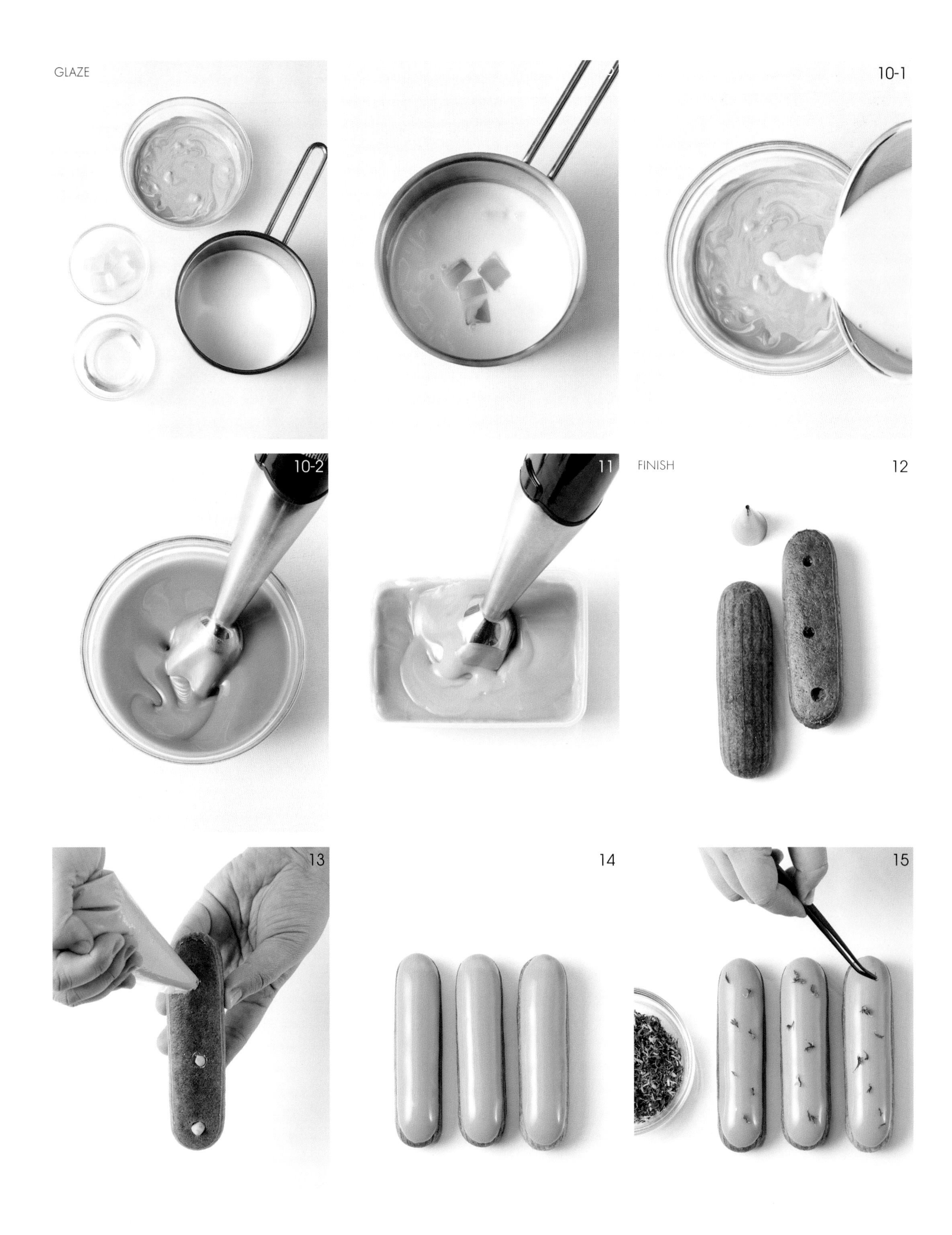
GLAZE
10-1
10-2
11
FINISH
12
13
14
15

글레이즈

9. 생크림과 물엿을 끓기 직전까지 가열한 후 젤라틴매스를 혼합한다.
10. 35℃로 녹인 블론드초콜릿과 코팅초콜릿에 **9**를 붓고 핸드블렌더로 혼합한다. 냉장고에서 12시간 세팅시킨다.
11. 사용할 때는 다시 30℃로 온도를 맞춘 다음 핸드블렌더로 균일하게 믹싱해 사용한다.

마무리

12. 구워낸 에클레어 바닥 부분에 작은 원형 깍지를 이용해 3개의 크림 주입구를 만든다.
13. 얼그레이 크림을 가득 채운 후 주입구를 깔끔하게 정리한다.
14. 에클레어 윗면을 글레이징한다.
15. 콘플라워를 올려 마무리한다.

GLAZE

9. Heat fresh cream and corn syrup until just before boiling, and stir in gelatin mass.
10. Pour the cream mixture(**9**) onto Blonde chocolate melted to 35°C and compound chocolate. Incorporate with a hand blender. Set in the fridge for 12 hours.
11. To use, reheat the glaze to 30°C, and mix thouroughly using a hand blender.

FINISH

12. Using a small round piping tip, make three holes on the bottom of the baked éclairs.
13. Fill with earl grey cream. Scrap off the excess around the holes.
14. Glaze the top of the éclairs.
15. Decorate with cornflowers to finish.

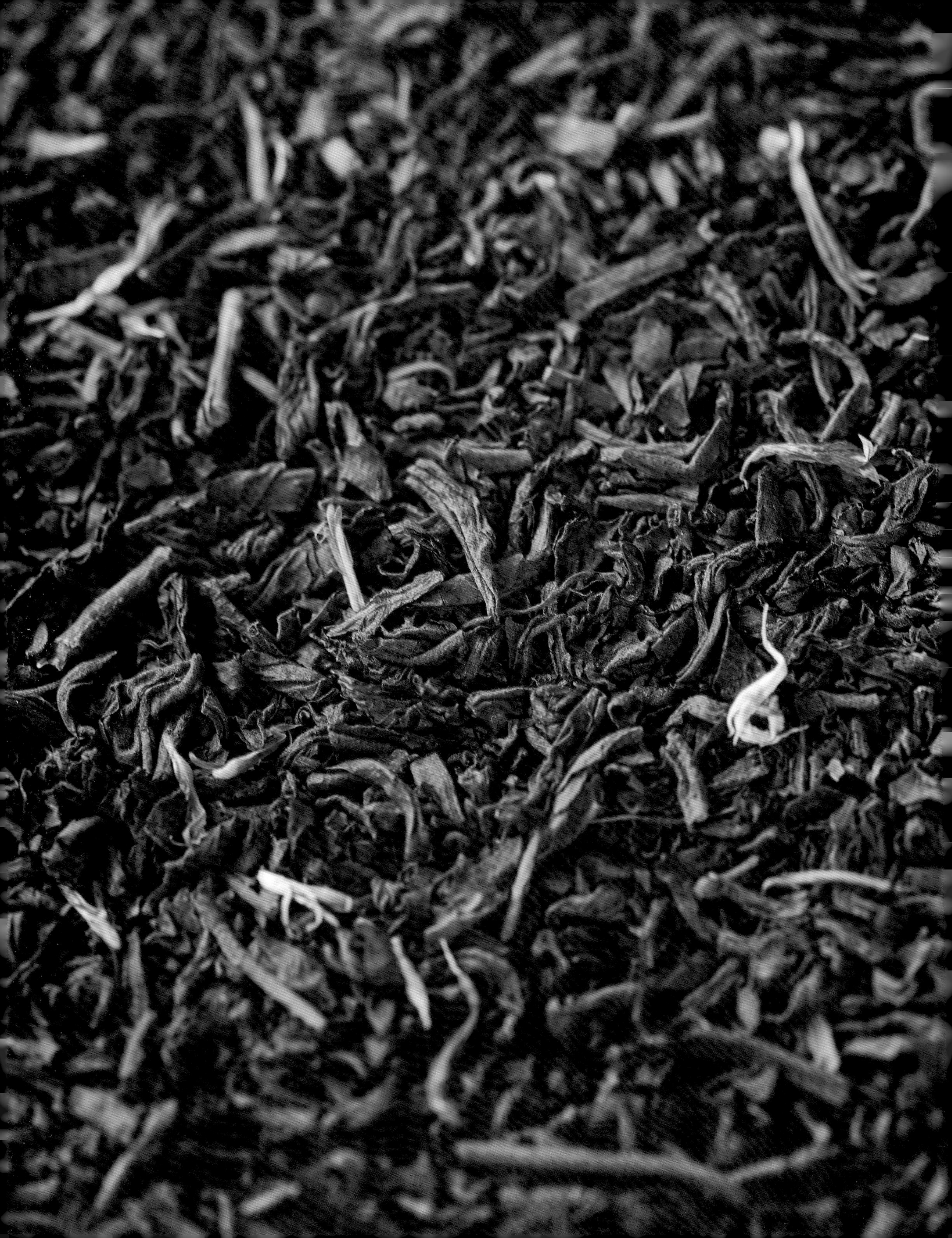

14 BROWN RICE GREEN TEA ECLAIR

현미녹차 에클레어

ingredients - 15ea

파트 아 슈

BASIC GLUTEN FREE

매실 젤

매실주 415g
유자주스 13g
설탕 66g
아가아가 6g

현미녹차 크림

달걀노른자 53g
설탕 74g
전분 24g
우유 412g
바닐라빈 1/2개
그린티파우더 2.5g
현미녹차 6g
버터 128g

글레이즈

생크림 183g
물엿 73g
젤라틴매스 42g
화이트초콜릿 231g
(OPALYS 33%)
화이트 코팅초콜릿 207g
그린티파우더 적당량

장식물

화이트초콜릿
(OPALYS 33%)
카카오버터
그린티파우더
현미녹차

PATE A CHOUX

BASIC GLUTEN FREE

MAESIL GEL

415g Maesil wine
13g Yuja juice
66g Sugar
6g Agar-agar

BROWN RICE GREEN TEA CREAM

53g Egg yolks
74g Sugar
24g Starch
412g Milk
1/2pc Vanilla bean
2.5g Green tea powder
6g Brown rice green tea
128g Butter

GLAZE

183g Fresh cream
73g Corn syrup
42g Gelatin mass
231g White chocolate
(OPALYS 33%)
207g White compound chocolate
Green tea powder QS

DECORATION

White chocolate
(OPALYS 33%)
Cacao butter
Green tea powder
Brown rice green tea

BROWN RICE GREEN TEA CREAM
1
2
3
4
5
6
7
8

Process

현미녹차 크림

1. 달걀노른자, 설탕 1/2, 전분을 혼합한다.
2. 냄비에 우유, 나머지 설탕, 바닐라빈, 그린티파우더를 넣고 끓기 직전까지 가열한 후 현미녹차를 넣고 향을 우린다.
3. **1**에 **2**를 조금씩 부어가며 혼합한다.
4. 체에 걸러 다시 냄비로 옮긴다.
5. 충분히 호화될 때까지 가열한다.
6. 45℃까지 식힌다.
7. 상온 상태의 버터를 넣고 핸드블렌더로 혼합한다.
8. 냉장고에서 12시간 휴지시켜 사용한다.

BROWN RICE GREEN TEA CREAM

1. Mix egg yolks, half of the sugar, and starch.
2. In a saucepan, heat milk, remaining sugar, vanilla bean, and green tea powder until just before boiling. Add brown rice green tea to infuse.
3. Gradually stir in the tea-infused mixture(**2**) into egg yolk mixture(**1**).
4. Strain into a saucepan.
5. Heat until gelatinized enough.
6. Let it cool to 45°C.
7. Mix with room temperature butter using a hand blender.
8. Rest in the fridge for 12 hours before use.

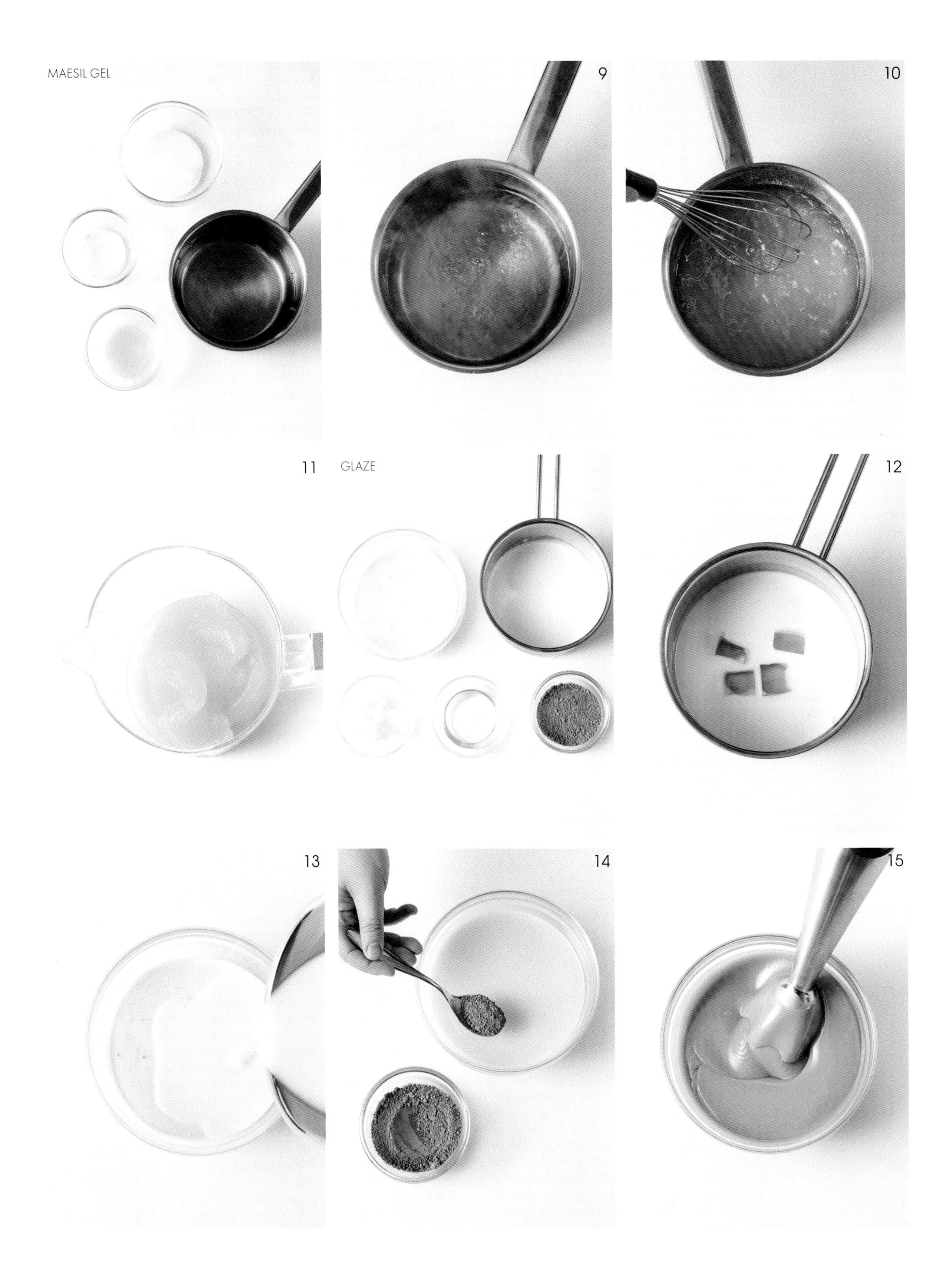
MAESIL GEL
9
10
11
GLAZE
12
13
14
15

매실주 젤

9. 냄비에 415g의 매실주를 넣고 알코올이 날아가도록 보글보글 끓인다. 매실주의 총량이 315g이 될 때까지 계속해서 가열한다.
10. 유자주스를 혼합한 후 미리 섞어둔 설탕과 아가아가를 넣고 끓어오를 때까지 가열한다. 볼에 옮겨 준 후 냉장고에서 굳혀준다.
11. 핸드블렌더를 이용해 부드러운 젤 상태가 될 때까지 갈아준다.

글레이즈

12. 생크림과 물엿을 끓기 직전까지 가열한 후 젤라틴매스를 혼합한다.
13. 35℃로 녹인 화이트초콜릿과 코팅초콜릿에 붓고 핸드블렌더로 혼합한다.
14. 그린티파우더를 넣고 계속해서 핸드블렌더로 혼합한 후 냉장고에서 12시간 세팅시킨다.
15. 사용할 때는 다시 30℃로 온도를 맞춘 다음 핸드블렌더로 균일하게 믹싱해 사용한다.

MAESIL GEL

9. In a saucepan, boil 415g of maesil wine until the alcohol evaporates. Continue to boil until it weighs 315g.
10. Mix yuja juice, and whisk in previously mixed sugar and agar-agar. Heat until the mixture boils. Transfer into a bowl and set in the fridge.
11. Blend using a hand blender until the mixture turns into a soft gel.

GLAZE

12. Heat fresh cream and corn syrup until just before boiling, and stir in gelatin mass.
13. Pour over white chocolate melted to 35°C and compound chocolate. Incorporate using a hand blender.
14. Add green tea powder, and continue to blend. Set in the fridge for 12 hours.
15. To use, reheat the glaze to 30°C, and mix thoroughly using a hand blender.

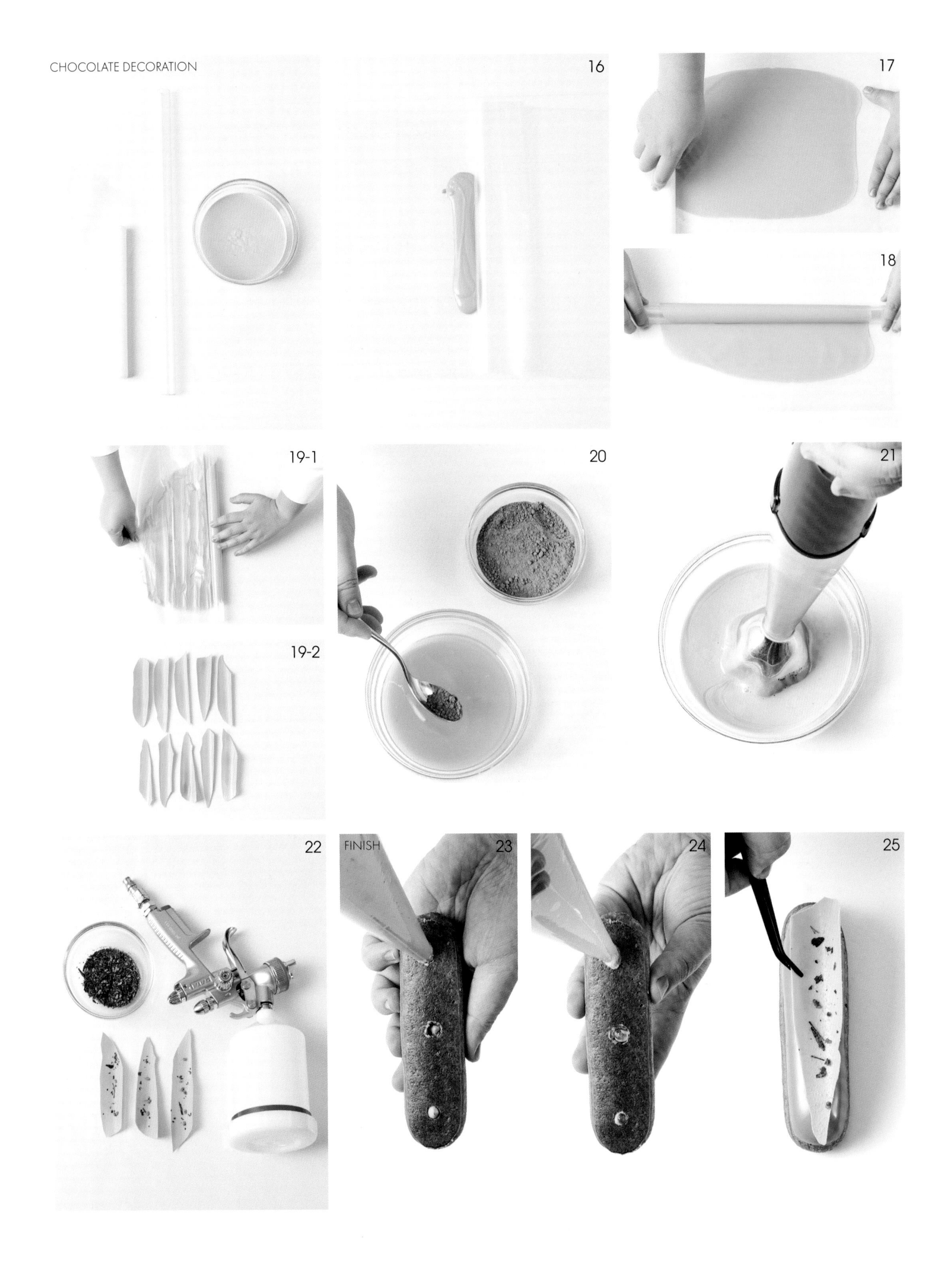
16
17
18
19-1
19-2
20
21
22
FINISH
23
24
25

초콜릿 장식물

16. 두 장의 투명 필름 사이에 그린티파우더를 섞어 템퍼링한 화이트초콜릿 적당량을 부어준다.
17. 필름을 덮고 밀대를 이용해 균일한 두께가 되도록 밀어편다.
18. 아크릴 파이프에 감아준다.
19. 충분히 세팅시킨 후 그대로 풀어준다. 자연스럽게 깨진 모양을 살려 필름에서 떼어낸 다음 냉동고에 30분간 넣어둔다.
20. 화이트초콜릿과 카카오버터를 1:1 비율로 섞고 그린티파우더를 첨가해 원하는 색으로 맞춘다.
21. 핸드블렌더로 혼합한다.
22. 차가워진 초콜릿 장식물 표면에 분사해 벨벳 같은 느낌을 표현한 후 현미녹차를 적당량 뿌려 마무리한다.

마무리

23. 구워낸 에클레어 바닥 부분에 작은 원형 깍지를 이용해 3개의 크림 주입구를 만든 후 현미녹차 크림을 90% 채운다.
24. 남은 공간에 매실주 젤을 가득 채운 후 주입구를 깔끔하게 정리한다.
25. 에클레어 윗면을 글레이징한 후 초콜릿 장식물을 올려 마무리한다.

CHOCOLATE DECORATION

16. Place a small amount of tempered white chocolate mixed with green tea powder, between two sheets of plastic film.
17. Roll into even thickness using a rolling pin.
18. Wrap around an acrylic pipe.
19. Let it crystallize completely, and unwrap as is. Let the chocolate break naturally, and remove the film carefully to save the shapes. Store in the freezer for 30 minutes.
20. Mix 1-part white chocolate with 1-part cacao butter, and add green tea powder to make desired color.
21. Incorporate with a hand blender.
22. Spray over the cold surface of the chocolate piece to give velvety texture, then sprinkle brown rice green tea.

FINISH

23. Using a small round piping tip, make three holes on the bottom of the baked éclairs. Fill with brown rice green tea cream about 90% of the cavity.
24. Pipe in maesil wine gel in the remaining space. Scrape off the excess around the holes.
25. Glaze the top of éclairs, and place chocolate decoration to finish.

15 ROSE ECLAIR

장미 에클레어

ingredients - 15ea

파트 아 슈

BASIC GLUTEN FREE

장식물

식용 장미

데코젤

라즈베리 리치 로즈 크림

달걀전란 182g

설탕 118g

라즈베리 리치 로즈 퓌레 148g

젤라틴매스 12.6g

버터 240g

라즈베리 리치 로즈 & 레몬 잼

라즈베리 퓌레 120g

라즈베리 리치 로즈 퓌레 120g

레몬주스 64g

설탕 80g

NH펙틴 4g

키르쉬 13g

글레이즈

생크림 183g

물엿 73g

젤라틴매스 42g

화이트초콜릿 231g
(OPALYS 33%)

화이트 코팅초콜릿 207g

붉은색 천연 식용 색소 적당량

PATE A CHOUX

BASIC GLUTEN FREE

DECORATION

Edible rose

DecoGel

RASPBERRY RICH ROSE CREAM

182g Whole eggs

118g Sugar

148g Raspberry Lychee Rose purée

12.6g Gelatin mass

240g Butter

RASPBERRY RICH ROSE & LEMON JAM

120g Raspberry purée

120g Raspberry Lychee Rose purée

64g Lemon juice

80g Sugar

4g Pectin NH

13g Kirsch

GLAZE

183g Fresh cream

73g Corn syrup

42g Gelatin mass

231g White chocolate
(OPALYS 33%)

207g White compound chocolate

Red food coloring (natural) QS

RASPBERRY RICH ROSE CREAM
1
2
3
4
BRAUN
RASPBERRY RICH ROSE & LEMON JAM
5
6
7

Process

라즈베리 리치 로즈 크림

1. 달걀전란과 설탕 1/2을 혼합한다. 라즈베리 리치 로즈 퓌레와 나머지 설탕을 45℃로 가열한 후 달걀 전란 믹스처에 조금씩 부어가며 혼합한다.
2. 냄비에 옮겨 78~80℃로 가열한 후 젤라틴매스를 넣어 섞고 체에 내려 볼에 옮겨준다.
3. 45℃까지 식힌 후 상온 상태의 버터를 넣고 핸드블렌더로 혼합한다.
4. 냉장고에서 12시간 휴지시킨 후 사용한다.

라즈베리 리치 로즈 & 레몬 잼

5. 냄비에 라즈베리 퓌레, 라즈베리 리치 로즈 퓌레, 레몬주스를 넣고 가열한 후 45℃가 되면 미리 섞어둔 설탕과 NH펙틴을 넣고 섞어준다.
6. 끓어오를 때까지 가열한 후 볼에 옮겨준다.
7. 키르쉬를 넣고 혼합해 완성한다.

RASPBERRY RICH ROSE CREAM

1. Mix whole eggs and half of the sugar. Heat raspberry lychee rose purée and remaining sugar to 45°C, and gradually stir into the whole egg mixture.
2. Transfer to a saucepan and heat until the temperature reaches 78~80°C. Stir in gelatin mass and strain into a bowl.
3. Cool down the mixture to 45°C, and mix with room temperature butter using a hand blender.
4. Rest in the fridge for 12 hours before use.

RASPBERRY RICH ROSE & LEMON JAM

5. In a saucepan, heat raspberry purée, raspberry lychee rose purée and lemon juice. When it reaches 45°C, whisk in previously mixed sugar and pectin NH.
6. Continue to heat until the mixture boils. Transfer to a bowl.
7. Whisk in Kirsch to finish.

GLAZE
8
9
10
11
FINISH
12
13
14
15

글레이즈

8. 끓기 직전까지 가열한 생크림과 물엿에 젤라틴매스를 혼합한 후 35℃로 녹인 화이트초콜릿과 코팅초콜릿에 부어준다.
9. 붉은색 천연 천연 식용 색소를 첨가한 후 핸드블렌더로 혼합한다.
10. 냉장고에서 12시간 세팅시킨다.
11. 사용할 때는 다시 30℃로 온도를 맞춘 다음 핸드블렌더로 균일하게 믹싱해 사용한다.

마무리

12. 구워낸 에클레어 바닥 부분에 작은 원형 깍지를 이용해 3개의 크림 주입구를 만든다. 라즈베리 리치 로즈 크림을 90% 채운다.
13. 남은 공간에 라즈베리 리치 로즈&레몬 잼을 가득 채운 후 주입구를 깔끔하게 정리한다.
14. 에클레어 윗면을 글레이징한다.
15. 식용 장미를 올리고 데코젤로 장식해 마무리한다.

GLAZE

8. Heat fresh cream and corn syrup until just before boiling, and pour onto white chocolate melted to 35°C and compound chocolate.
9. Add red natural food coloring and mix using a hand blender.
10. Set in the fridge for 12 hours.
11. To use, reheat the glaze to 30°C, and mix thoroughly using a hand blender.

FINISH

12. Using a small round piping tip, make three holes on the bottom of the baked éclairs. Fill with raspberry lychee rose cream about 90% of the cavity.
13. Pipe in raspberry lychee rose & lemon jam in the remaining space. Scrape off the excess around the holes.
14. Glaze the top of the éclairs.
15. Place edible rose petals and decorate with decogel to finish.

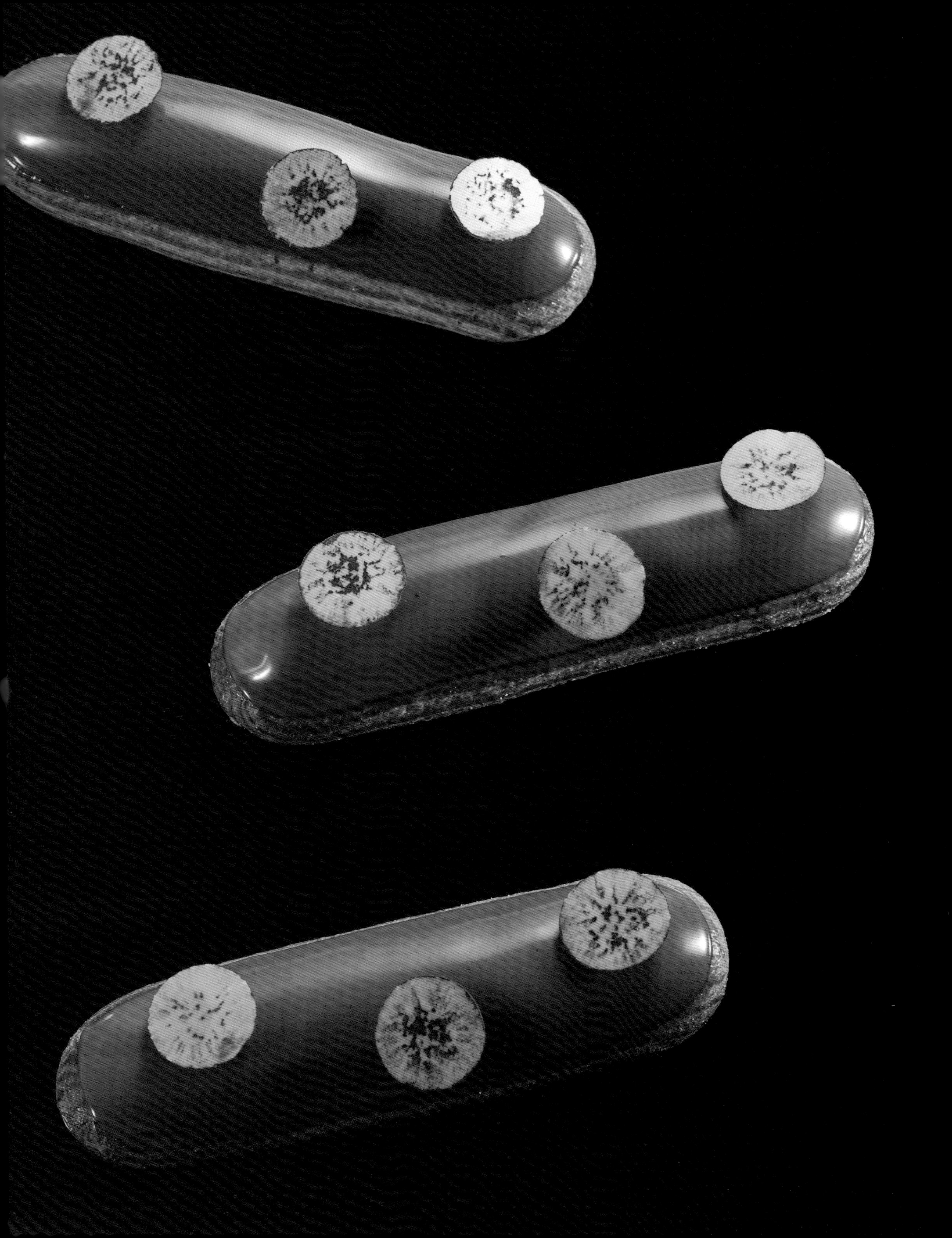

16 COFFEE ECLAIR

커피 에클레어

ingredients - 15ea

파트 아 슈

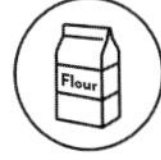

BASIC GLUTEN FREE

커피 크림

우유 402g
원두 14g
달걀노른자 52g
설탕 74g
전분 24g
버터 126g
커피농축액 9g

글레이즈

원두 10g
생크림 207g
물엿 69g
젤라틴매스 42g
밀크초콜릿 220g
(JIVARA LATTE 40%)
화이트 코팅초콜릿 197g

장식물

블론드초콜릿
(DULCEY 32%)
카카오파우더
(POUDRE DE CACAO)

PATE A CHOUX

BASIC GLUTEN FREE

COFFEE CREAM

402g Milk
14g Coffee beans
52g Egg yolks
74g Sugar
24g Starch
126g Butter
9g Coffee extract

GLAZE

10g Coffee beans
207g Fresh cream
69g Corn syrup
42g Gelatin mass
220g Milk chocolate
(JIVARA LATTE 40%)
197g White compound chocolate

DECORATION

Blonde chocolate
(DULCEY 32%)
Cacao powder
(POUDRE DE CACAO)

COFFEE CREAM
1
2
3
4
5
6
7
8

Process

커피 크림

1. 원두는 160℃ 오븐에서 5분간 데워 준비한다. 냄비에 우유와 설탕 1/2을 넣고 끓기 직전까지 가열한 후 준비한 원두를 넣는다.
2. 핸드블렌더를 이용해 원두를 잘게 분쇄한 후 향을 우린다.
3. 달걀노른자, 나머지 설탕, 전분을 혼합한다.
4. **3**에 **2**를 조금씩 부어가며 섞어준다.
5. 체에 걸러 냄비로 옮긴다.
6. 충분히 호화될 때까지 가열한다.
7. 볼에 옮겨준 후 커피농축액을 섞고 45℃까지 식혀준다.
8. 상온 상태의 버터를 넣고 핸드블렌더로 혼합한 후 냉장고에서 12시간 휴지시켜 사용한다.

COFFEE CREAM

1. Warm coffee beans in 160°C oven for 5 minutes. In a saucepan, heat milk and half of the sugar until just before boiling, and add prepared coffee beans.
2. Grind the beans using a hand blender, and let them infuse.
3. In a bowl, mix egg yolks, remaining sugar, and starch.
4. Gradually stir in the egg mixture(**3**) into the infused mixture(**2**).
5. Strain into a saucepan.
6. Heat until gelatinized enough.
7. Pour into a bowl and mix with coffee extract. Cool to 45°C.
8. Mix with room temperature butter using a hand blender, and rest in the fridge for 12 hours before use.

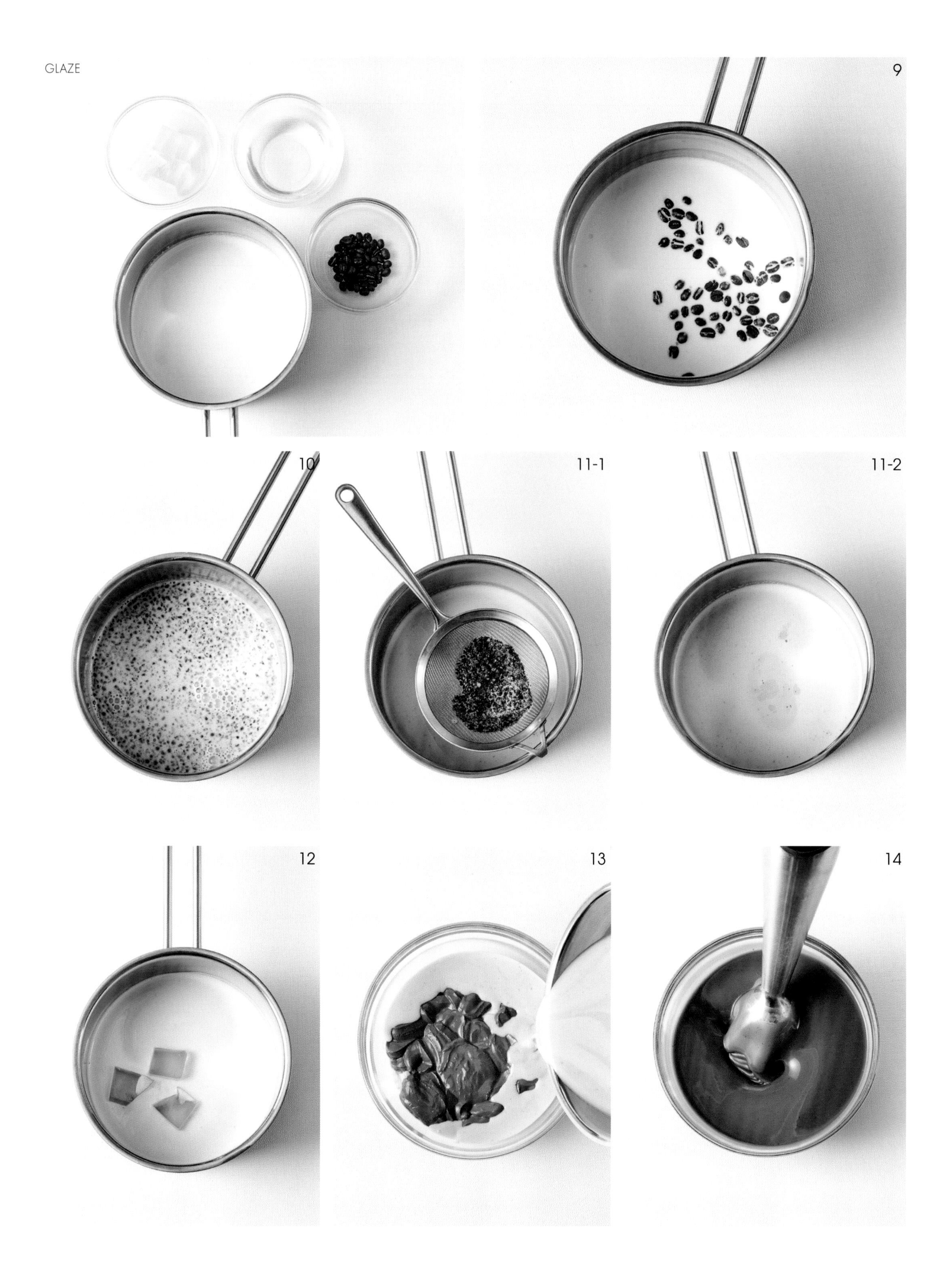
9
10
11-1
11-2
12
13
14

글레이즈

9. 끓기 직전까지 가열한 생크림과 물엿에 160℃ 오븐에서 5분간 데운 원두를 넣는다.
10. 핸드블렌더로 원두를 잘게 분쇄한 후 향을 우린다.
11. 체에 걸러준다.
12. 젤라틴매스를 혼합한다.
13. 35℃로 녹인 밀크초콜릿과 코팅초콜릿에 붓고 핸드블렌더로 혼합한다. 냉장고에서 12시간 세팅시킨다.
14. 사용할 때는 다시 30℃로 온도를 맞춘 다음 핸드블렌더로 균일하게 믹싱해 사용한다.

GLAZE

9. Heat fresh cream and corn syrup until just before boiling. Add coffee beans warmed in 160°C oven for 5 minutes.
10. Grind the beans using a hand blender, and let them infuse.
11. Strain the mixture.
12. Stir in gelatin mass.
13. Pour over milk chocolate melted to 35°C and compound chocolate. Incorporate using a hand blender. Set in the fridge for 12 hours.
14. To use, reheat the glaze to 30°C, and mix thouroughly using a hand blender.

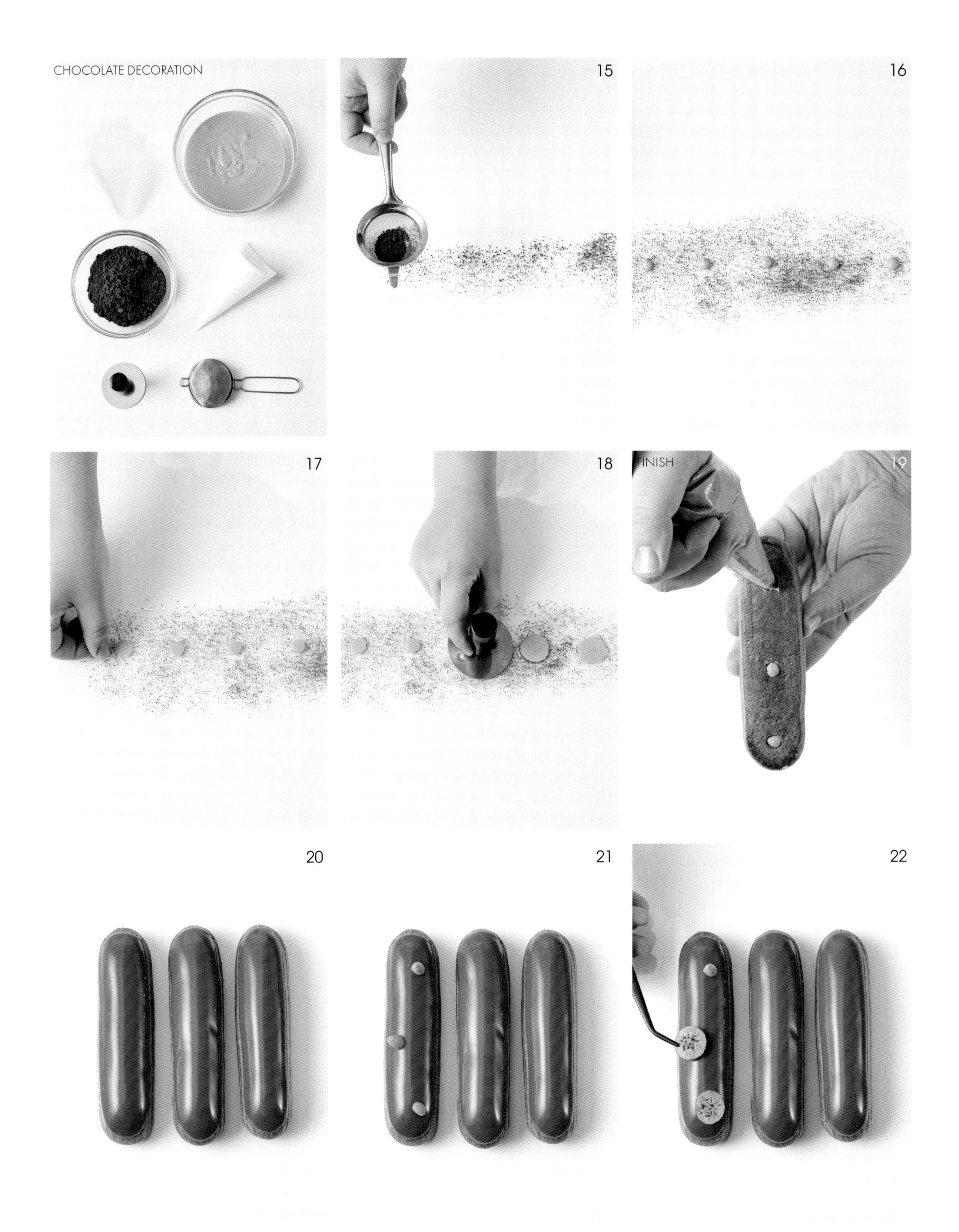
CHOCOLATE DECORATION
15
16
17
18
FINISH
19
20
21
22

초콜릿 장식물

15. 투명한 필름 위에 카카오파우더를 적당량 뿌려준다.
16. 템퍼링한 블론드초콜릿을 파이핑한다.
17. 파이핑한 블론드초콜릿 위에 다시 필름을 덮어준다.
18. 평평한 도구를 이용해 천천히 눌러 초콜릿이 자연스럽게 퍼지게 한다.

마무리

19. 구워낸 에클레어 바닥 부분에 작은 원형 깍지를 이용해 3개의 크림 주입구를 만든 후 커피 크림을 가득 채운다.
20. 주입구를 깔끔하게 정리한 후 에클레어 윗면을 글레이징한다.
21. 에클레어 윗면에 여분의 커피 크림을 파이핑한다.
22. 초콜릿 장식물을 올려 마무리한다.

CHOCOLATE DECORATION

15. Sprinkle cacao powder on clear plastic film.
16. Pipe tempered Blonde chocolate on top.
17. Cover with another clear plastic film over the piped Blonde chocolate.
18. Carefully press the surface using a flat tool to make the chocolate spread out naturally.

FINISH

19. Using a small round piping tip, make three holes on the bottom of the baked éclairs. Fill it with coffee cream.
20. Scrap off the excess and glaze the top of the éclair.
21. Pipe some of the remaining coffee cream on top.
22. Place the chocolate decorations to finish.

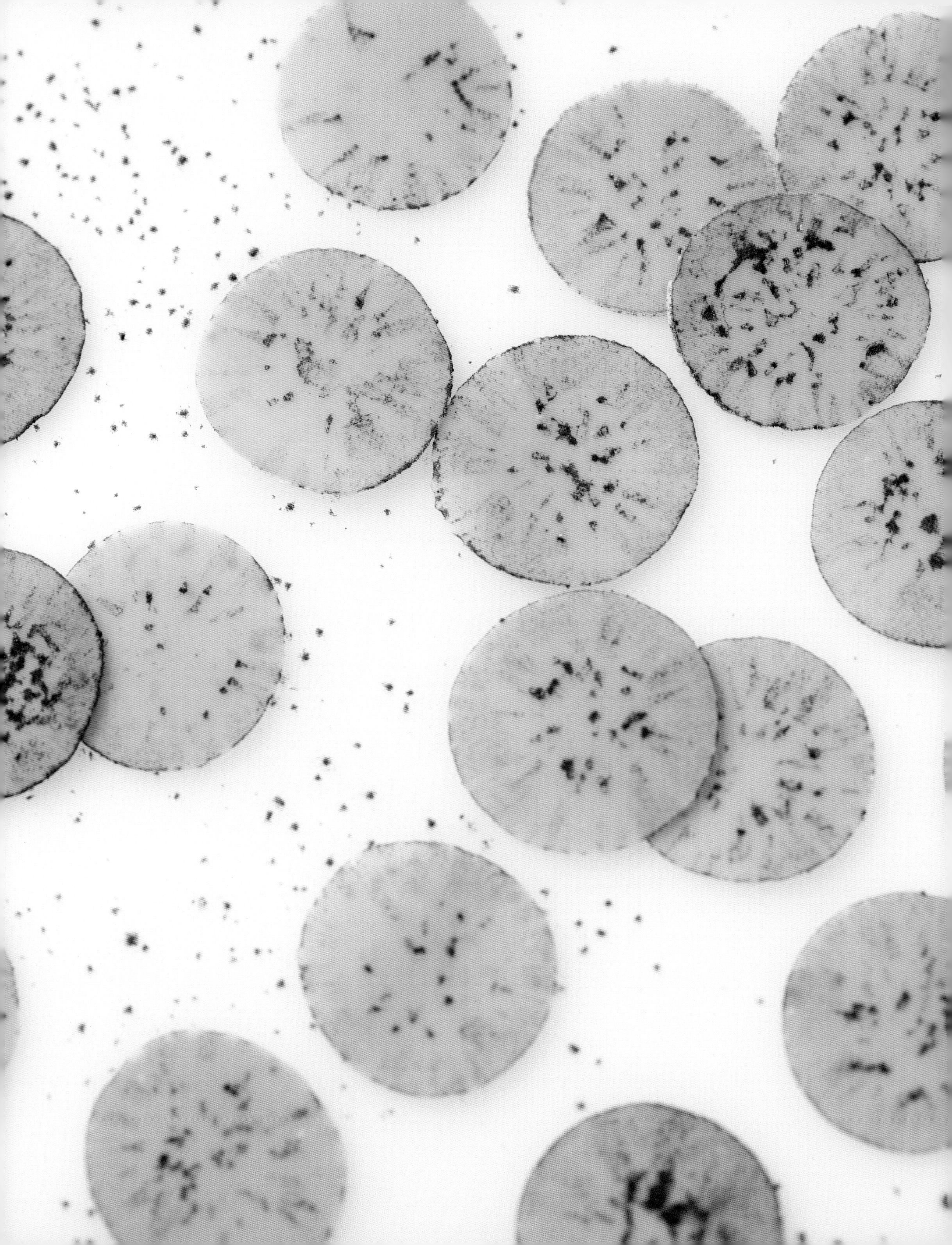

17 SALTED BUTTER CARAMEL ECLAIR

솔티드 버터 캐러멜 에클레어

ingredients - 15ea

파트 아 슈

BASIC GLUTEN FREE

장식물

무염버터
게랑드소금

캐러멜 크림

설탕 60g
물엿 40g
물 33g
소금 0.7g
우유 400g
바닐라빈 1/2개
달걀노른자 48g
전분 22g
버터 116g

캐러멜 소스

물엿 99g
설탕 99g
생크림 151g
바닐라빈 1/3개
소금 1.6g
젤라틴매스 4.8g
버터 49g

글레이즈

설탕 33g
생크림 212g
물엿 45g
젤라틴매스 42g
캐러멜 밀크초콜릿 213g
(CARAMELIA 36%)
화이트 코팅초콜릿 190g

PATE A CHOUX

BASIC GLUTEN FREE

DECORATION

Unsalted butter
Guerande salt

CARAMEL CREAM

60g Sugar
40g Corn syrup
33g Water
0.7g Salt
400g Milk
1/2pc Vanilla bean
48g Egg yolks
22g Starch
116g Butter

CARAMEL SAUCE

99g Corn syrup
99g Sugar
151g Fresh cream
1/3pc Vanilla bean
1.6g Salt
4.8g Gelatin mass
49g Butter

GLAZE

33g Sugar
212g Fresh cream
45g Corn syrup
4.2g Gelatin mass
213g Caramel milk chocolate
(CARAMELIA 36%)
190g White compound chocolate

CARAMEL CREAM
1
2
3
4-1
4-2
5
6
7

Process

캐러멜 크림

1. 달걀노른자와 전분을 혼합한다.
2. 우유에 바닐라빈과 소금을 넣고 60℃로 데운다.
3. 냄비에 물엿과 설탕을 넣고 캐러멜라이즈한다. 끓기 직전까지 가열한 물을 넣고 섞어준 후 80℃까지 식힌다.
4. 60℃로 데운 우유를 **3**에 부어 혼합한 후 **1**에 조금씩 넣어가며 섞는다.
5. 체에 걸러 다시 냄비로 옮겨준 후 충분히 호화될 때까지 가열한다.
6. 45℃까지 식힌 후 상온 상태의 버터를 넣고 핸드블렌더로 혼합한다.
7. 냉장고에서 12시간 휴지시킨 후 사용한다.

CARAMEL CREAM

1. Mix egg yolks and starch.
2. Add vanilla bean and salt in milk and heat to 60°C.
3. In a saucepan, caramelize corn syrup and sugar together. Stir in water heated until just before boiling, and cool it down to 80°C.
4. Stir in milk heated to 60°C into caramel mixture(**3**). Gradually incorporate to egg yolk mixture(**1**).
5. Strain into a saucepan and heat until gelatinized enough.
6. Cool to 45°C and mix in room temperature butter using a hand blender.
7. Rest in the fridge for 12 hours before use.

CARAMEL SAUCE
8
9
10
11
12
GLAZE
13
14
15

캐러멜 소스

8. 생크림에 바닐라빈과 소금을 넣고 끓기 직전까지 가열한다.
9. 냄비에 물엿과 설탕을 넣고 가열해 캐러멜라이즈한다.
10. **9**에 **8**을 붓고 섞어준 후 108℃까지 가열한다.
11. 60℃까지 식힌 후 젤라틴매스를 넣고 섞는다.
12. 상온 상태의 버터를 넣고 핸드블렌더로 혼합한 후 상온에서 12시간 휴지시켜 사용한다.

글레이즈

13. 냄비에 설탕을 넣고 캐러멜라이즈한다. 생크림과 물엿은 끓기 직전까지 가열한다.
14. 캐러멜라이즈한 설탕에 데운 생크림과 물엿을 넣고 섞어준 후 60℃까지 식힌다.
15. 젤라틴매스를 넣고 섞은 다음 체에 내려 35℃로 녹인 캐러멜 밀크초콜릿과 코팅초콜릿에 부어준다. 핸드블렌더를 이용해 혼합한 후 냉장고에서 12시간 세팅시킨다.
16. 사용할 때는 다시 30℃로 온도를 맞춘 다음 핸드블렌더로 균일하게 믹싱해 사용한다.

CARAMEL SAUCE

8. Add vanilla bean and salt to fresh cream and heat until just before boiling.
9. In a saucepan, caramelize corn syrup and sugar together.
10. Stir in cream mixture(**8**) into caramel(**9**), and cook until 108°C.
11. Cool down to 60°C, and stir in gelatin mass.
12. Incorporate with room temperature butter using a hand blender, and rest in ambient temperature for 12 hours before use.

GLAZE

13. In a saucepan, caramelize the sugar. In a separate saucepan, heat fresh cream and corn syrup until just before boiling.
14. Stir in the heated cream mixture into caramelized sugar, and cool to 60°C.
15. Stir in gelatin mass, and strain over caramel chocolate melted to 35°C and compound chocolate. Incorporate with a hand blender, and set in the fridge for 12 hours.
16. To use, reheat the glaze to 30°C, and mix thoroughly using a hand blender.

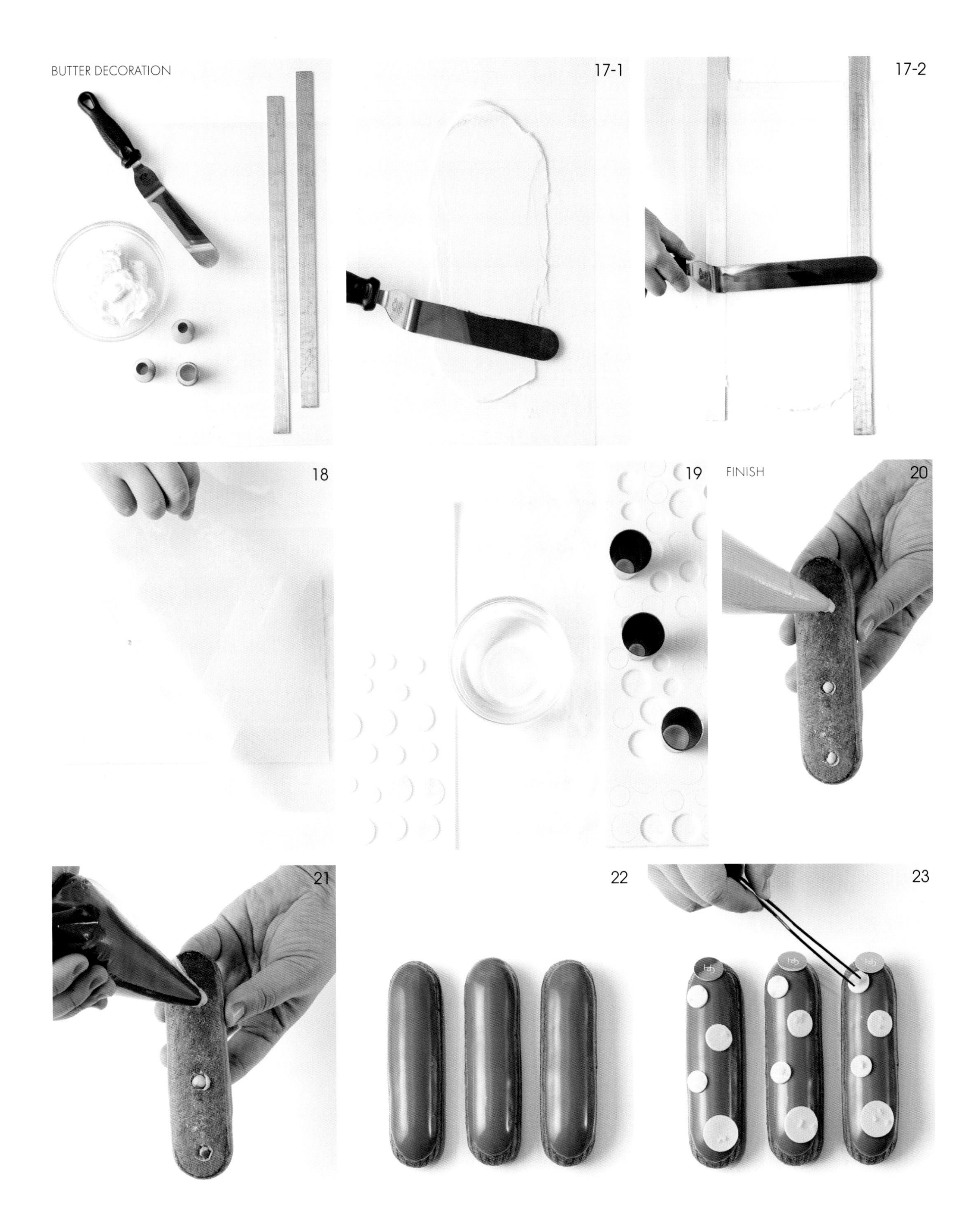
BUTTER DECORATION
17-1
17-2
18
19
FINISH
20
21
22
23

버터 장식물

17. 텍스처 시트에 상온 상태의 버터를 2mm 두께로 펼친 후 냉장고에서 굳혀준다.
18. 버터가 단단하게 굳으면 시트를 벗겨낸다.
19. 따뜻한 물을 이용해 원형 커터를 데워준 후 버터를 커팅한다.

마무리

20. 구워낸 에클레어 바닥 부분에 작은 원형 깍지를 이용해 3개의 크림 주입구를 만든 후 캐러멜 크림을 90% 채운다.
21. 남은 공간에 캐러멜 소스를 가득 채운 후 주입구를 깔끔하게 정리한다.
22. 에클레어 윗면을 글레이징한다.
23. 버터 장식물과 게랑드소금을 올려 마무리한다.

BUTTER DECORATION

17. Spread room temperature butter on a textured sheet in 2mm thickness and store in the fridge to set.
18. Remove the sheet when the butter hardens.
19. Cut the butter using a round cutter warmed with warm water.

FINISH

20. Using a small round piping tip, make three holes on the bottom of the baked éclairs. Fill with caramel cream about 90% of the cavity.
21. Fill in caramel sauce in the remaining space. Scrape off the excess around the holes.
22. Glaze the top of éclairs.
23. Place butter decorations and Guerande salt to finish.

Cream & Milk

CREAM BRULEE ECLAIR

YOGURT ECLAIR

TIRAMISU ECLAIR

GARUHARU

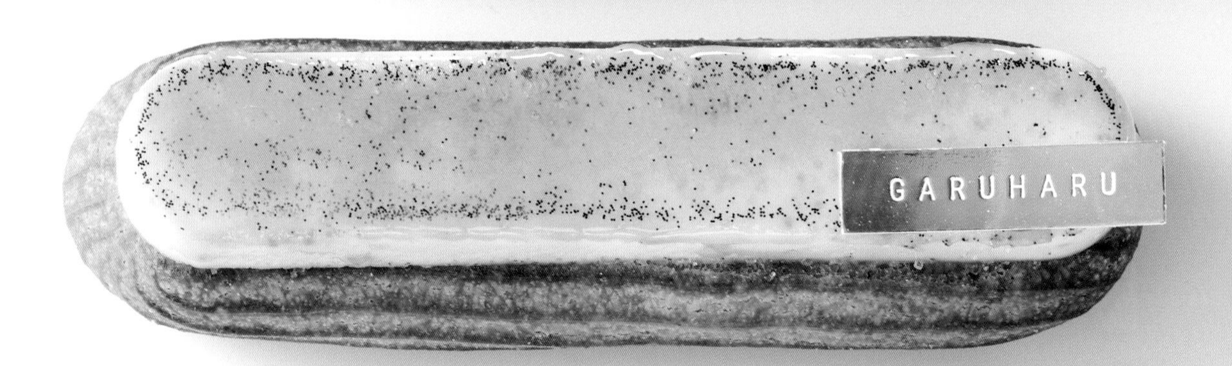
GARUHARU

GARUHARU

18 CREAM BRULEE ECLAIR

크림 브륄레 에클레어

ingredients - 15ea

파트 아 슈

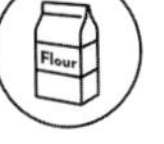

캐러멜 소스(217p)

물엿 99g
설탕 99g
생크림 151g
바닐라빈 1/3개
소금 1.6g
젤라틴매스 4.8g
버터 49g

크림 브륄레

설탕 60g
달걀노른자 60g
달걀흰자 60g
생크림 300g
럼 20g
바닐라빈 2/3개

더블 바닐라 크림

달걀노른자 53g
설탕 75g
전분 24g
우유 419g
타히티 바닐라빈 2/3개
마다가스카르 바닐라빈 2/3개
버터 129g

PATE A CHOUX

CARAMEL SAUCE(217p)

99g Corn syrup
99g Sugar
151g Fresh cream
1/3pc Vanilla bean
1.6g Salt
4.8g Gelatin mass
49g Butter

CREAM BRULEE

60g Sugar
60g Egg yolks
60g Egg whites
300g Fresh cream
20g Rum
2/3pc Vanilla bean

DOUBLE VANILLA CREAM

53g Egg yolks
75g Sugar
24g Starch
419g Milk
2/3pc Tahiti vanilla bean
2/3pc Madagascar vanilla bean
129g Butter

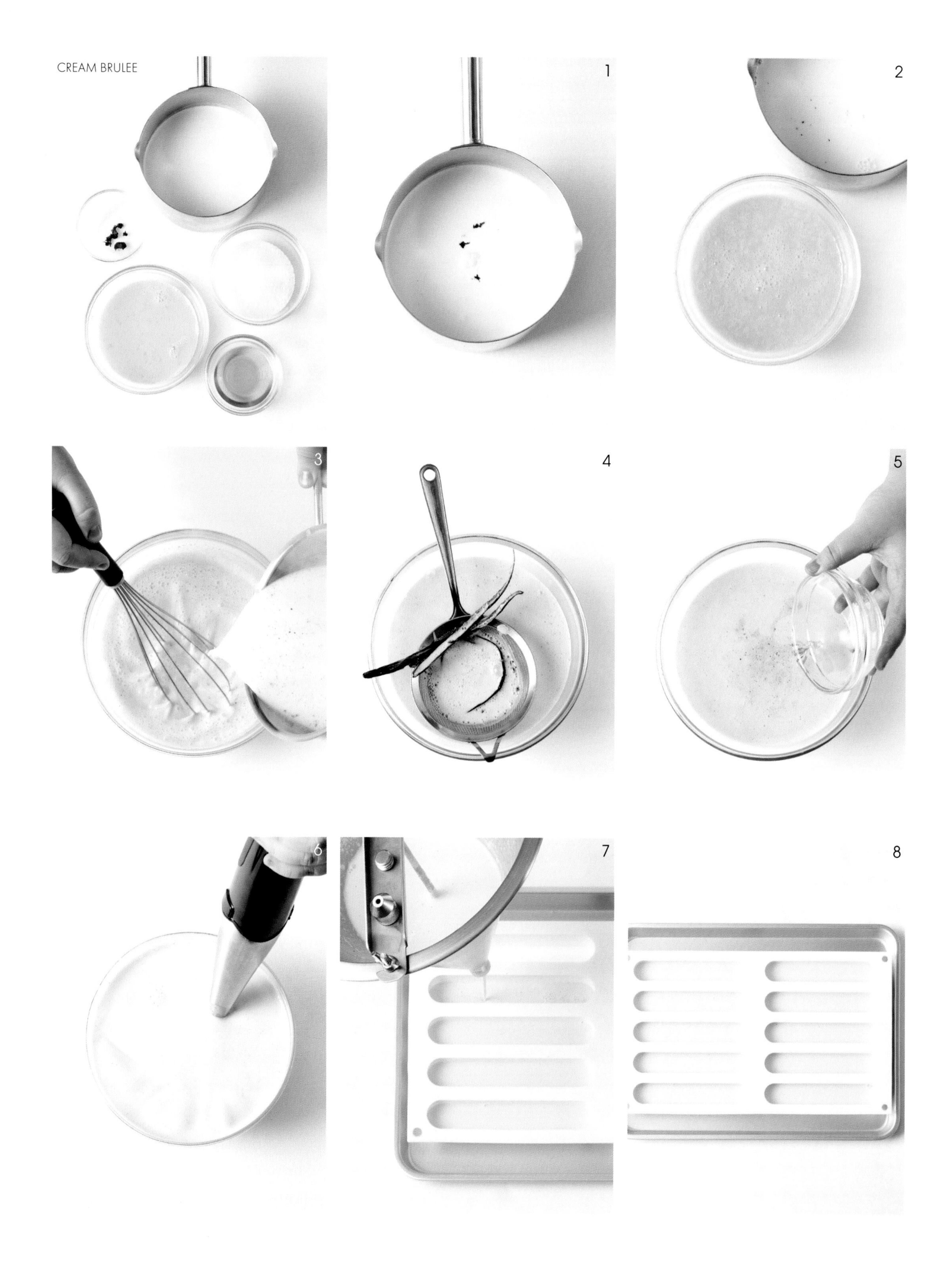
CREAM BRULEE
1
2
3
4
5
6
7
8

Process

크림 브륄레

1. 냄비에 생크림, 설탕 1/2, 바닐라빈을 넣고 80℃로 데운 후 향을 우린다.
2. 볼에 달걀노른자, 달걀흰자, 나머지 설탕을 넣고 섞는다.
3. 바닐라 향이 충분히 우러나면 **2**에 **1**을 조금씩 부어가며 섞는다.
4. 체에 걸러준다.
5. 럼을 넣고 섞는다.
6. 핸드블렌더로 혼합한 후 냉장고에서 12시간 휴지시킨다.
7. 실리콘 몰드(에클레어 모양)에 40g씩 채운다.
8. 80℃로 예열된 오븐에서 약 60분간 구워준 후 냉동고에서 단단하게 굳힌다.

CREAM BRULEE

1. In a saucepan, heat fresh cream, half of the sugar, and vanilla bean to 80°C, and let it infuse.
2. In a bowl, mix egg yolks, egg whites, and remaining sugar.
3. When the aroma of the vanilla has been extracted enough, gradually stir in the egg mixture(**2**) into the vanilla mixture(**1**).
4. Strain the mixture.
5. Add rum.
6. Incorporate using a hand blender, and let it rest in the fridge for 12 hours.
7. Fill 40g each in silicon mold (éclair shape).
8. Bake in an oven preheated to 80°C for about 60 minutes. Freeze completely.

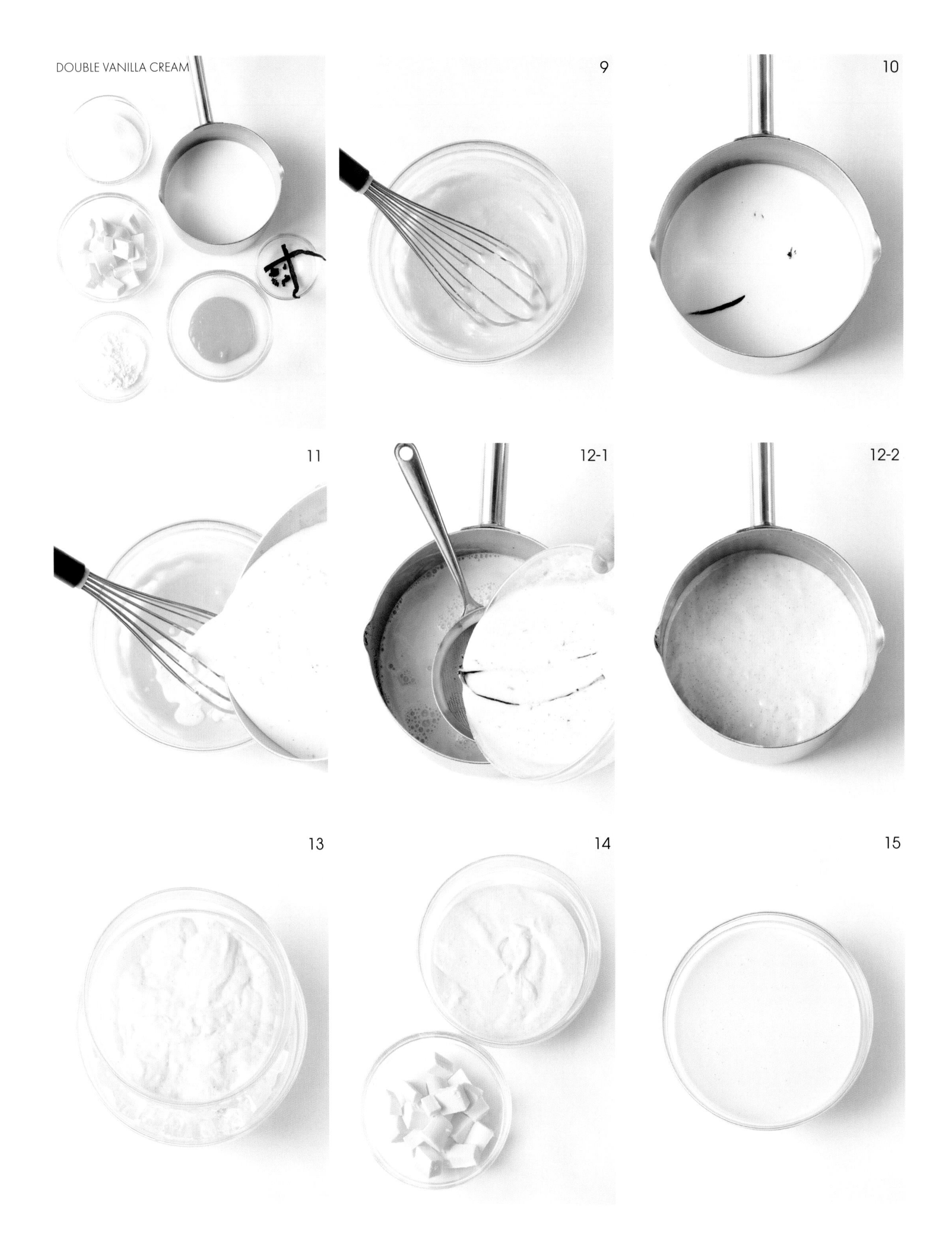
9
10
11
12-1
12-2
13
14
15

더블 바닐라 크림

9. 달걀노른자와 설탕 1/2, 전분을 혼합한다.
10. 냄비에 우유, 나머지 설탕, 두 종류의 바닐라빈을 넣고 끓기 직전까지 가열한 후 향을 우린다.
11. **9**에 **10**을 조금씩 부어가며 혼합한다.
12. 체에 걸러 다시 냄비에 옮겨준 후 충분히 호화될 때까지 가열한다.
13. 45℃까지 식힌다.
14. 상온 상태의 버터를 넣고 핸드블렌더로 혼합한다.
15. 냉장고에서 12시간 휴지시킨 후 사용한다.

DOUBLE VANILLA CREAM

9. Mix egg yolks, half of the sugar, and starch.
10. In a saucepan, heat milk, remaining sugar, and two types of vanilla beans until just before boiling, and let it infuse.
11. Stir in the vanilla-infused mixture(**10**) into egg yolks mixture(**9**), a little bit at a time.
12. Strain into a saucepan and heat until gelatinized enough.
13. Cool down to 45°C.
14. Incorporate with room temperature butter using a hand blender.
15. Rest in the fridge for 12 hours before use.

FINISH

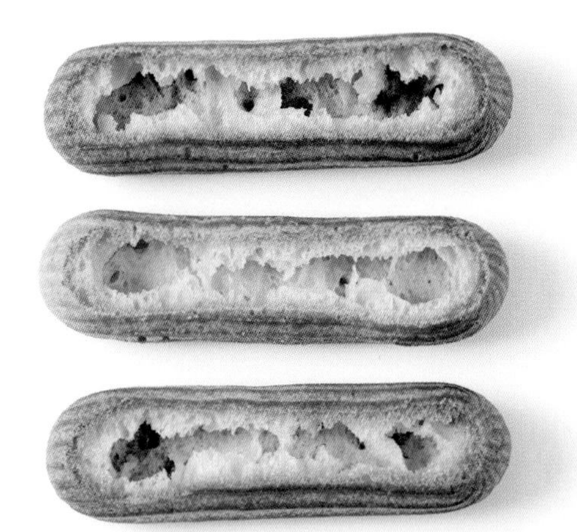

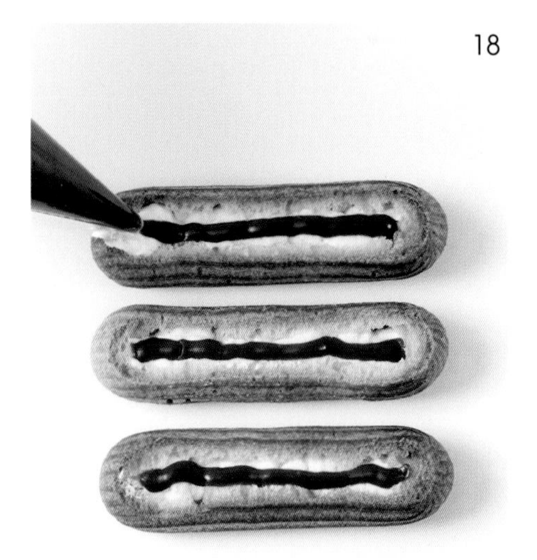

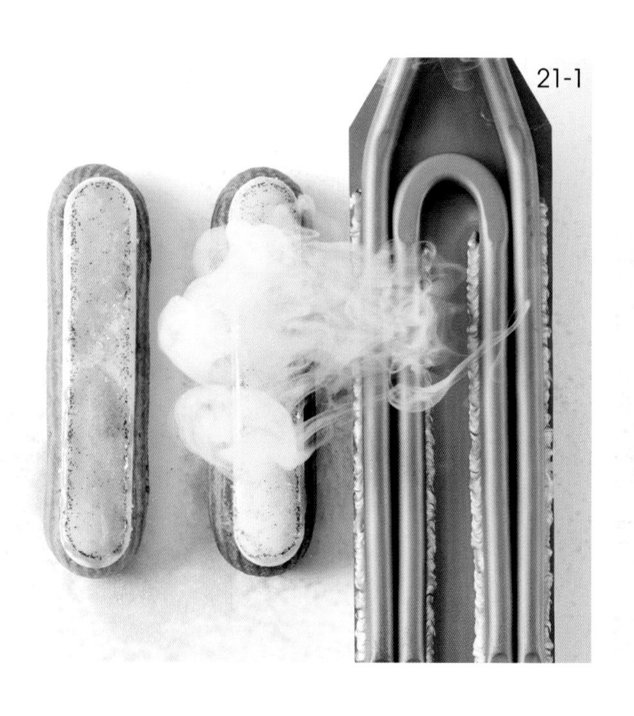

마무리

16. 구워낸 에클레어의 윗부분을 제거한다.
17. 더블 바닐라 크림을 채운다.
18. 캐러멜 소스를 적당량 파이핑한다.
19. 냉동고에서 단단하게 굳힌 크림 브륄레를 얹어준다.
20. 에클레어 윗면에 카소나드를 충분히 뿌려준다.
21. 카라멜라이저 또는 토치를 이용해 카소나드를 캐러멜라이징한다.

FINISH

16. Cut the top of the baked éclairs.
17. Fill with double vanilla cream.
18. Pipe in a moderate amount of caramel sauce.
19. Place completely frozen cream brulee on top.
20. Sprinkle cassonade generously on top of the éclairs.
21. Caramelize cassonade using caramelizer or blow torch.

19 YOGURT ECLAIR

요거트 에클레어

ingredients - 15ea

파트 아 슈	요거트 크림	글레이즈	슈거 페이스트*	장식물
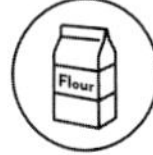 BASIC / GLUTEN FREE	달걀노른자 51g	생크림 183g	슈거파우더 250g	슈거 페이스트*
	설탕 75g	물엿 73g	레몬즙 5g	코코넛 분말
	전분 24g	젤라틴매스 42g	젤라틴매스 24g	
	요거트파우더(sosa) 21g	화이트초콜릿 231g (OPALYS 33%)		
	우유 404g	화이트 코팅초콜릿 207g		
	버터 126g			

PATE A CHOUX	YOGURT CREAM	GLAZE	SUGAR PASTE*	DECORATION
  BASIC / GLUTEN FREE	51g Egg yolks	183g Fresh cream	250g Powdered sugar	Sugar paste*
	75g Sugar	73g Corn syrup	5g Lemon juice	Coconut powder
	24g Starch	42g Gelatin mass	24g Gelatin mass	
	21g Yogurt powder (Sosa)	231g White chocolate (OPALYS 33%)		
	404g Milk	207g White compound chocolate		
	126g Butter			

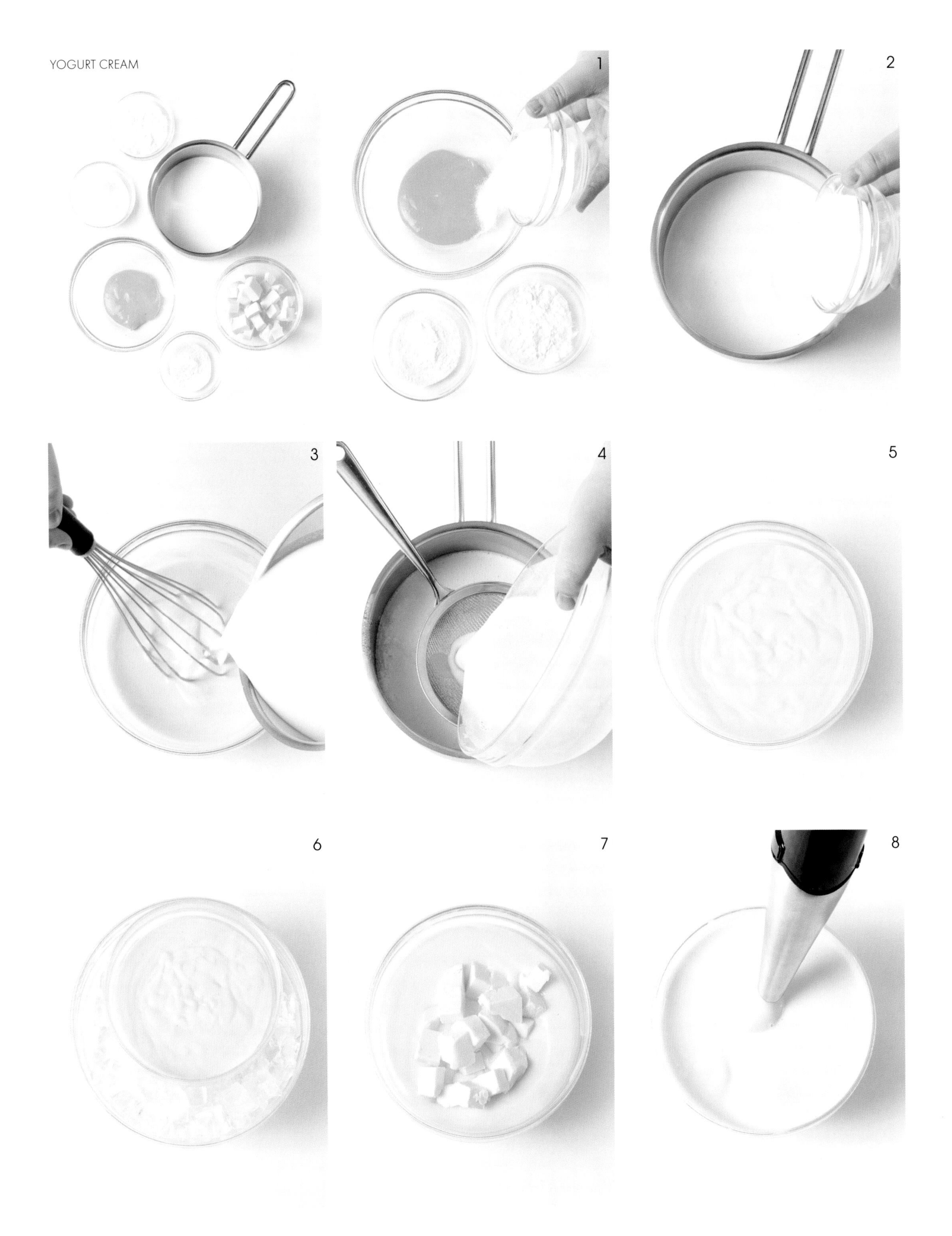
1
2
3
4
5
6
7
8

Process

요거트 크림

1. 달걀노른자, 설탕 1/2, 전분, 요거트파우더를 섞는다.
2. 냄비에 우유와 나머지 설탕을 넣고 45℃로 가열한다.
3. **1**에 **2**를 조금씩 부어가며 혼합한다.
4. 체에 걸러 다시 냄비로 옮겨준다.
5. 충분히 호화될 때까지 가열한 후 볼에 옮겨준다.
6. 45℃까지 식혀준다.
7. 상온 상태의 버터를 넣고 핸드블렌더로 혼합한다.
8. 냉장고에서 12시간 휴지시킨 후 사용한다.

YOGURT CREAM

1. Mix egg yolks, half of sugar, starch, and yogurt powder.
2. In a saucepan, heat milk and remaining sugar to 45°C.
3. Stir in milk mixture(**2**) little by little into egg yolk mixture(**1**).
4. Strain and pour back into a saucepan.
5. Heat until gelatinized enough, and transfer into a bowl.
6. Cool down to 45°C.
7. Incorporate with room temperature butter using a hand blender.
8. Rest in the fridge for 12 hours before use.

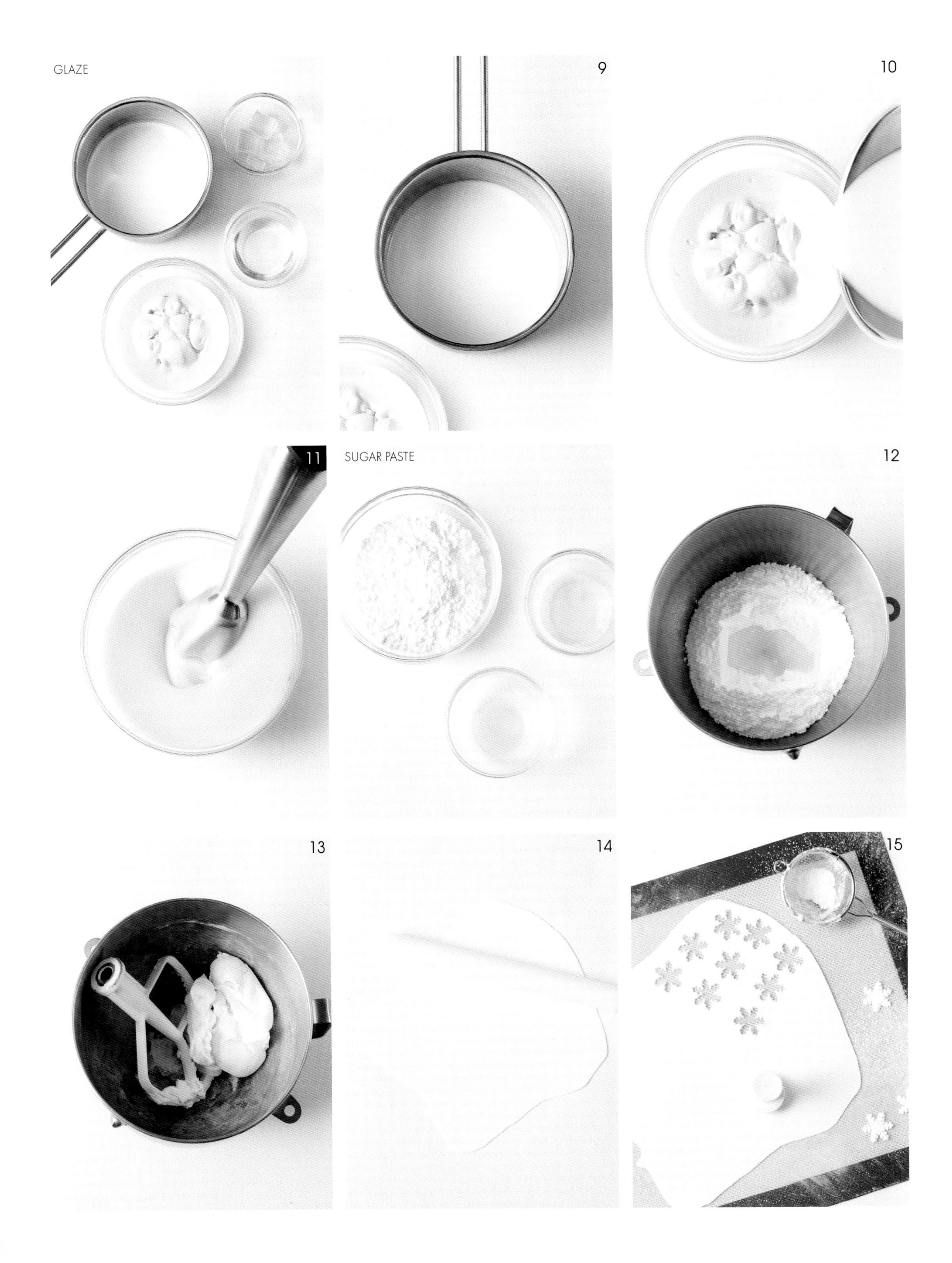
GLAZE
9
10
11
SUGAR PASTE
12
13
14
15

글레이즈

9. 냄비에 생크림과 물엿을 넣고 끓기 직전까지 가열한 후 젤라틴매스를 혼합한다.
10. 35℃로 녹인 화이트 초콜릿과 코팅초콜릿에 부어준다. 핸드블렌더로 혼합한 후 냉장고에서 12시간 세팅시킨다.
11. 사용할 때는 다시 30℃로 온도를 맞춘 다음 핸드블렌더로 균일하게 믹싱해 사용한다.

슈거 페이스트

12. 슈거파우더, 레몬즙, 녹인 젤라틴매스를 믹싱볼에 담아준다.
13. 반죽이 한 덩어리가 될 때까지 믹싱한다.
14. 반죽을 최대한 얇게 밀어 편다.
15. 원하는 모양의 커터로 커팅한 후 상온에서 하루 동안 건조시켜 사용한다.

GLAZE

9. In a saucepan, heat fresh cream and corn syrup until just before boiling, and stir in gelatin mass.
10. Pour over white chocolate melted to 35°C and compound chocolate. Incorporate with a hand blender. Set in the fridge for 12 hours.
11. To use, reheat the glaze to 30°C, and mix thoroughly using a hand blender.

SUGAR PASTE

12. Combine powdered sugar, lemon juice, and melted gelatin mass in a mixing bowl.
13. Mix until the mixture forms one dough.
14. Roll out the dough as thin as possible.
15. Cut into desired shapes using cutters, and dry at ambient temperature for one day.

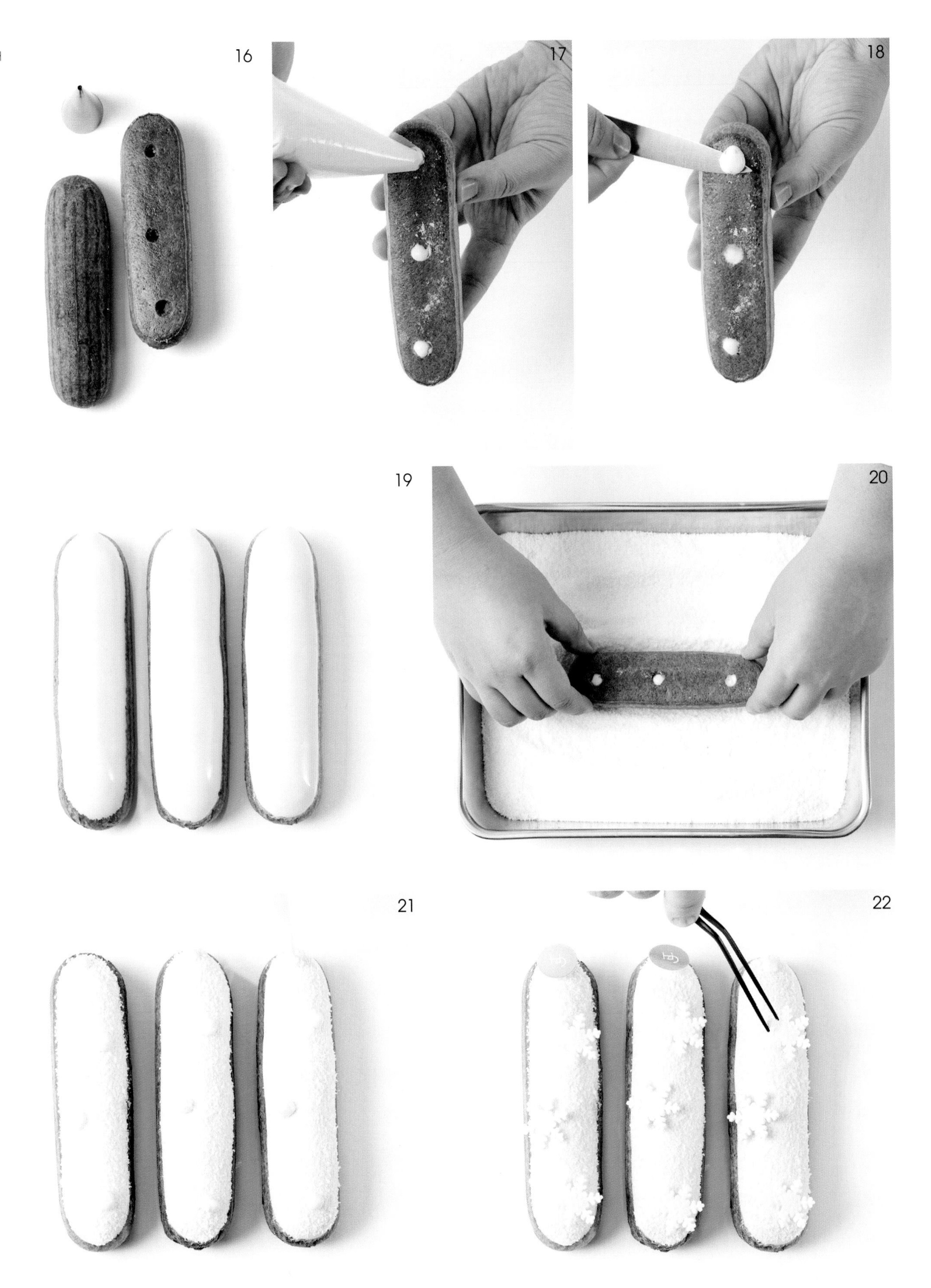
16
17
18
19
20
21
22

마무리

16. 구워낸 에클레어 바닥 부분에 작은 원형 깍지를 이용해 3개의 크림 주입구를 만든다.
17. 요거트 크림을 가득 채운다.
18. 주입구를 깔끔하게 정리한다.
19. 에클레어 윗면을 글레이징한다.
20. 글레이즈가 굳기 전 코코넛 분말을 듬뿍 묻혀준다.
21. 장식물을 올릴 위치에 여분의 글레이즈를 파이핑한다.
22. 슈거 페이스트 장식물을 올려 마무리한다.

FINISH

16. Using a small round piping tip, make three holes on the bottom of the baked éclairs.
17. Fill with yogurt cream.
18. Scrape off the excess around the holes.
19. Glaze the top of éclairs.
20. Press on coconut powder before the glaze sets.
21. Pipe dots with remaining glaze on desired spots.
22. Place the sugar paste decorations to finish.

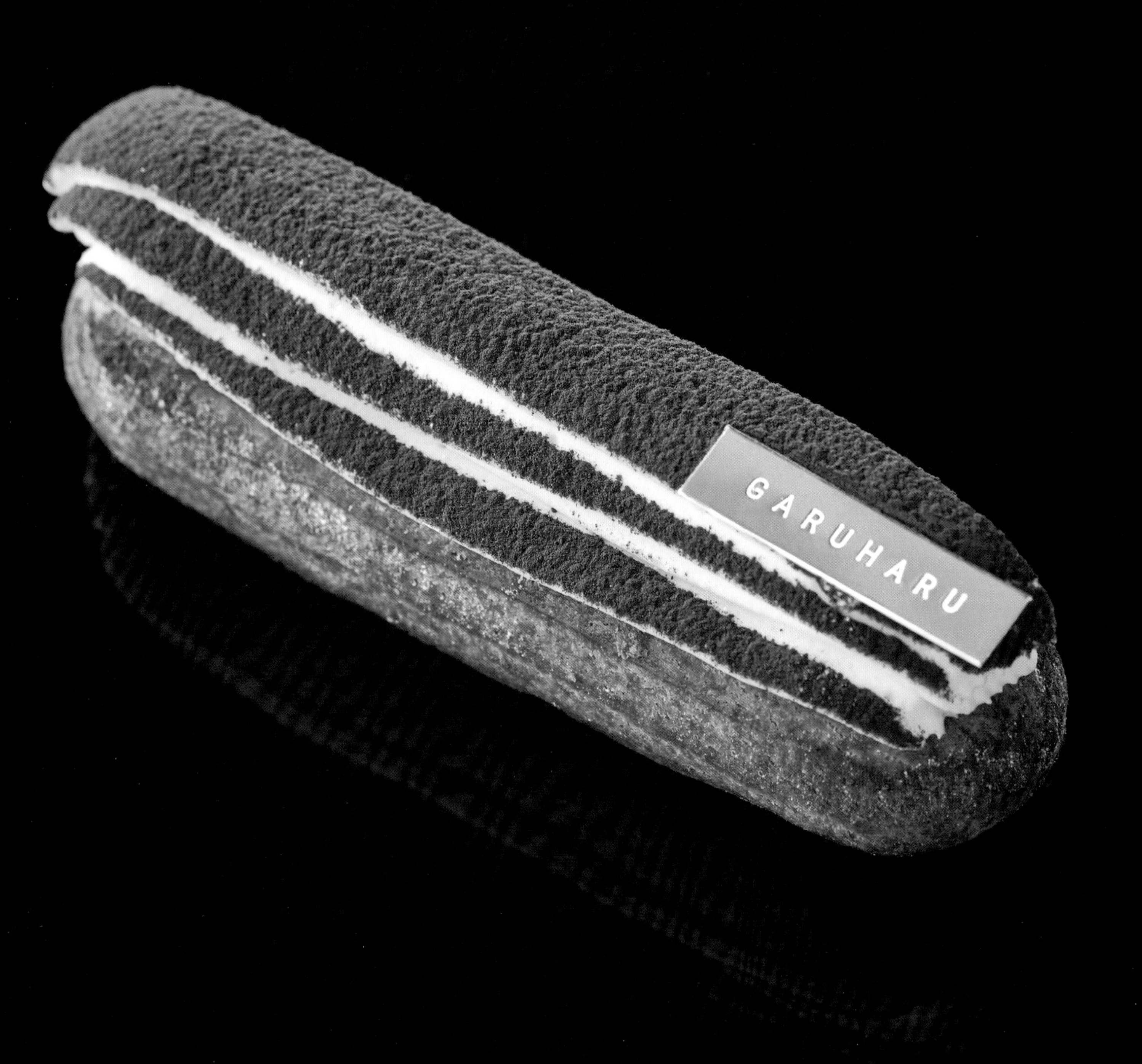
GARUHARU

20 TIRAMISU ECLAIR

티라미수 에클레어

ingredients - 15ea

파트 아 슈

BASIC

커피 크림(205p) 701g

마스카르포네 크림

달걀노른자 40g
설탕 22g
생크림 55g
우유 55g
마스카르포네치즈 500g

커피 시럽

30보메 시럽(물 1000g, 설탕 1350g) 120g
에스프레소 240g

기타

SAVOIARDI
(＊글루텐 함유)
데코스노우
카카오파우더

PATE A CHOUX

BASIC

COFFEE CREAM(205p)
701g

MASCARPONE CREAM

40g Egg yolks
22g Sugar
55g Fresh cream
55g Milk
500g Mascarpone cheese

COFFEE SYRUP

120g 30 Baume syrup (water 1000g, Sugar 1350g)
240g Espresso

ETC

SAVOIARDI
(*contains gluten)
Decosnow
Cacao powder

MASCARPONE CREAM
1
2
3
4-1
4-2
83.0
ON / OFF
5
6
7

Process

마스카르포네 크림

1. 달걀노른자와 설탕 1/2을 혼합한다.
2. 냄비에 생크림, 우유, 나머지 설탕을 넣고 45℃로 가열한다.
3. **1**에 **2**를 조금씩 나눠 넣으면서 섞는다.
4. 냄비에 옮겨 83~85℃로 가열해 크렘 앙글레이즈를 만든다.
5. 체에 걸러준 후 8℃로 차갑게 식힌다.
6. 마스카르포네치즈와 혼합한다.
7. 사용하기 직전 휘핑해 사용한다.

MASCARPONE CREAM

1. Mix egg yolks and half of the sugar.
2. In a saucepan, heat fresh cream, milk, and and remaining sugar to 45°C.
3. Gradually stir in cream mixture(**2**) into egg yolk mixture(**1**), a little bit at a time.
4. Pour into a saucepan, and cook until 83~85°C to make crème anglais.
5. Strain and cool down to 8°C.
6. Incorporate with mascarpone cheese.
7. Whip just before use.

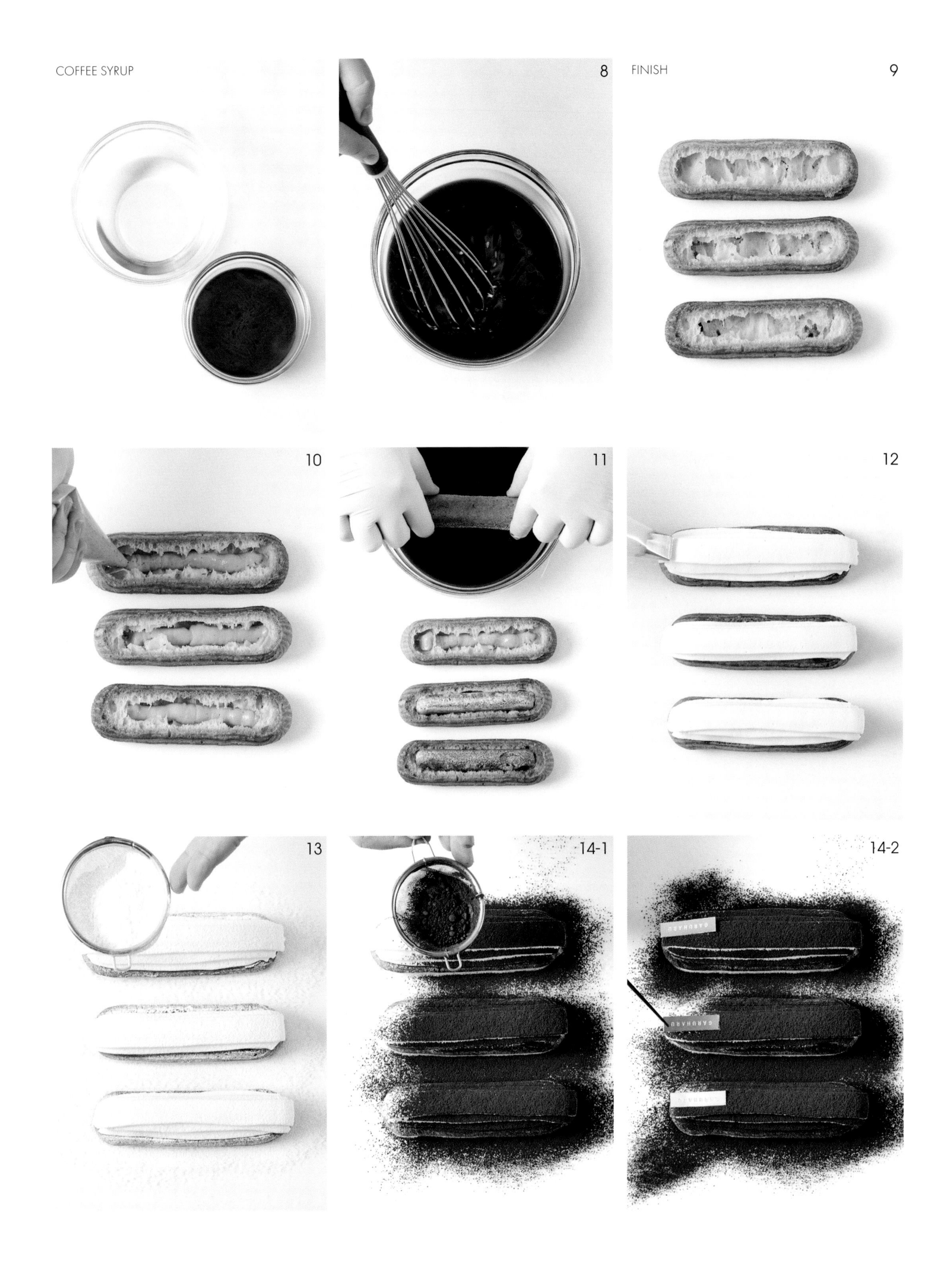
COFFEE SYRUP
8
FINISH
9
10
11
12
13
14-1
14-2

커피 시럽

8. 30보메 시럽과 에스프레소를 혼합한다.

마무리

9. 구워낸 에클레어의 윗부분을 제거한다.
10. 커피 크림을 20g씩 채운다.
11. 커피 시럽에 흠뻑 적신 SAVOIARDI 쿠키를 채운 후 표면을 고르게 정리한다.
12. 895번 깍지를 이용해 마스카르포네 크림을 파이핑한다.
13. 데코스노우를 뿌린다. (데코스노우는 카카오파우더가 크림의 수분을 흡수하는 것을 막아준다.)
14. 카카오파우더를 뿌려 마무리한다.

COFFEE SYRUP

8. Combine 30 Baume syrup and espresso.

FINISH

9. Remove the top of the baked éclair.
10. Pipe 20g of coffee cream.
11. Fill SAVOIARDI cookies soaked in coffee syrup and even out the surface.
12. Pipe mascarpone cream using a Basketweave nozzle (#895).
13. Dust with decosnow. (Decosnow helps prevent cacao powder from absorbing moisture in the cream.)
14. Dust with cacao powder to finish.

Special Recipe

CHOUX A LA CREME

CHOUQUETTE

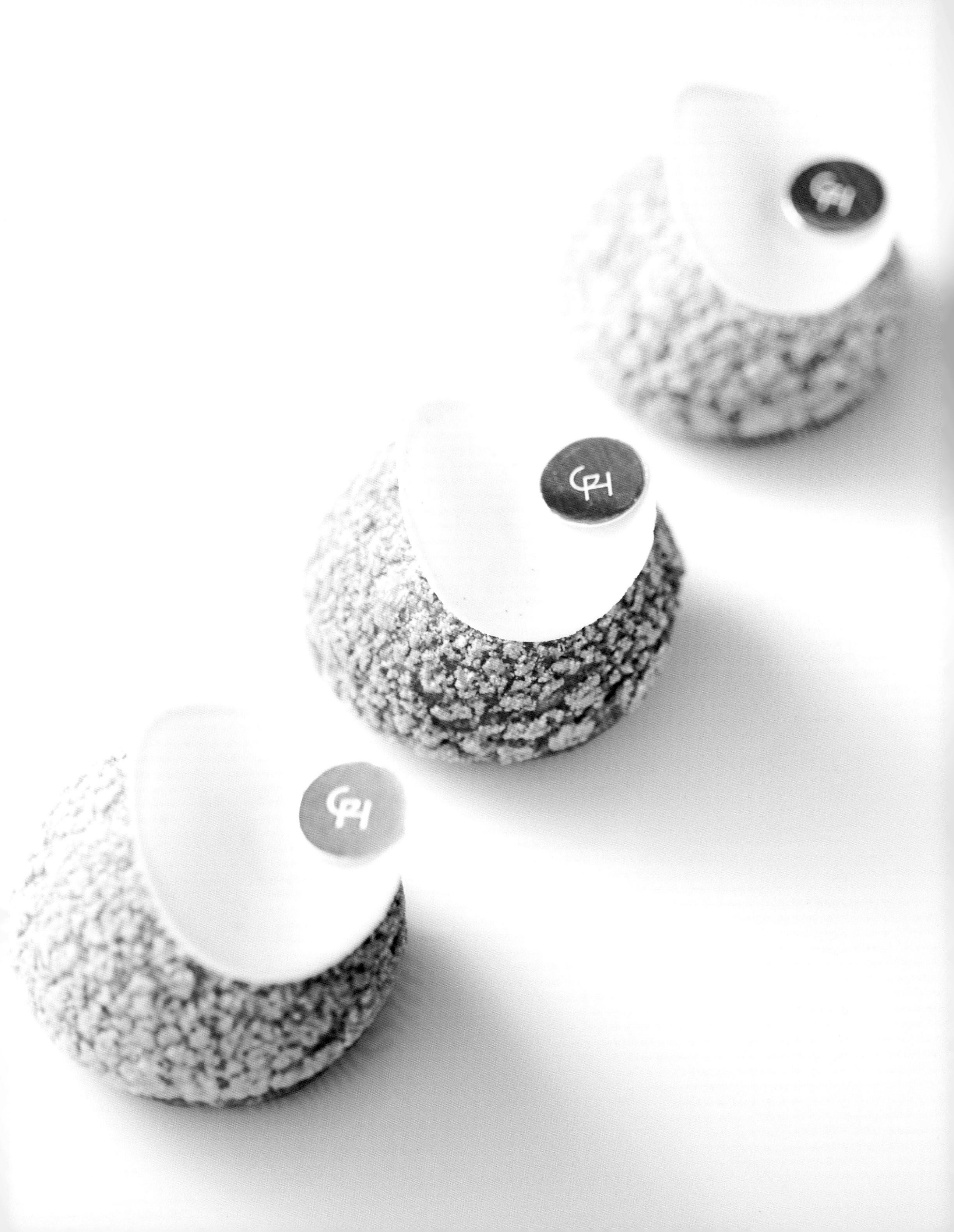

21 CHOUX A LA CREME

슈 아 라 크렘

ingredients - 15ea

파트 아 슈

BASIC GLUTEN FREE

장식물

화이트초콜릿
(OPALYS 33%)

바닐라빈

스트로이젤(BASIC)

버터 100g
박력분 100g
아몬드파우더(발렌시아) 25g
설탕 125g

스트로이젤 (GLUTEN FREE)

버터 100g
쌀가루 100g
아몬드파우더(발렌시아) 25g
설탕 125g

크렘 파티시에

설탕 131g
달걀노른자 126g
전분 42g
우유 526g
바닐라빈 1개
버터 42g

크렘 레제

크렘 파티시에 600g
생크림 150g
설탕 15g

PATE A CHOUX

BASIC GLUTEN FREE

DECORATION

White chocolate
(OPALYS 33%)

Vanilla bean

STREUSEL(BASIC)

100g Butter
100g Cake flour
25g Almond powder (Valencia)
125g Sugar

STREUSEL (GLUTEN FREE)

100g Butter
100g Rice flour
25g Almond powder (Valencia)
125g Sugar

CREME PATISSIER

131g Sugar
126g Egg yolks
42g Starch
526g Milk
1pc Vanilla bean
42g Butter

CREME LEGERE

600g Crème patissiere
150g Fresh cream
15g Sugar

STREUSEL(BASIC)
1
2
Heavy Duty
3
Heavy Duty
4
5
6
FINISH
7
8

Process

스트로이젤 (BASIC)

1. 버터를 부드럽게 풀어준다.
2. 설탕을 넣고 믹싱한다.
3. 박력분과 아몬드파우더를 넣고 믹싱한다.
4. 한 덩어리가 되면 반죽을 마무리한다.
5. 밀대를 이용해 2mm 두께로 밀어편다.
6. 냉동고에 잠시 두어 단단하게 굳힌 후 지름 4.8cm 원형 커터를 이용해 커팅한다.

TIP. 스트로이젤(GLUTEN FREE)

❶ 버터를 부드럽게 풀어준다.
❷ 설탕을 넣고 믹싱한다.
❸ 습식 멥쌀가루와 아몬드파우더를 넣고 믹싱한다.
❹ 한 덩어리가 되면 반죽을 마무리한다.

마무리

7. 원형(지름 1cm) 깍지를 이용해 지름 4.8cm 원형으로 슈 반죽을 파이핑한다.
8. 슈 반죽 위에 커팅한 스트로이젤을 올려준 후 160℃에서 30분간 굽는다. (컨벡션 오븐 기준)

STREUSEL (BASIC)

1. Using a paddle attachment, beat the butter to soften.
2. Add sugar and continue mixing.
3. Mix in cake flour and almond powder.
4. Remove when the dough becomes homogeneous.
5. Roll out to 2mm thinkness.
6. Keep in a freezer until firm. Use a 4.8cm diameter round cutter to cut.

TIP. Streusel (GLUTEN FREE)

❶ Soften butter.
❷ Add sugar, mix.
❸ Mix in wet-milled nonglutinous rice flour and almond powder.
❹ Remove when the dough becomes homogeneous.

FINISH

7. Pipe choux paste using 1cm round nozzle into 4.8cm diameter round shape.
8. Place previously cut streusel disks on top of the piped choux paste. Bake in an oven set to 160°C for 30 minutes (convection oven).

CREME PATISSIERE
9
10
11
12
13
14
15
CREME LEGERE
16

크렘 파티시에

9. 냄비에 우유, 설탕 1/2, 바닐라빈을 넣고 끓기 직전까지 가열한 후 향을 우린다.
10. 볼에 달걀노른자, 나머지 설탕, 전분을 넣고 섞어준다.
11. **10**에 **9**를 조금씩 부어가며 섞는다.
12. 체로 걸러 다시 냄비로 옮겨준다.
13. 충분히 호화시킨다.
14. 상온 상태의 버터를 넣고 혼합한다.
15. 빠르게 식혀준 후 냉장보관한다.

크렘 레제

16. 크렘 파티시에는 부드럽게 풀어준다. 생크림과 설탕을 섞어 단단하게 거품을 올린 다음 크렘 파티시에와 혼합해 크렘 레제를 완성한다.

CREME PATISSIERE

9. In a saucepan, heat milk, half of the sugar, and vanilla bean until just before boiling.
10. In a bowl, mix egg yolks, remaining sugar, and starch.
11. Gradually stir in milk mixture(**9**) into egg yolk mixture(**10**).
12. Strain into a saucepan.
13. Heat until gelatinized enough.
14. Incorporate with room temperature butter.
15. Cool rapidly and store in the fridge.

CREME LEGERE

16. Soften crème patissiere. Whip fresh cream and sugar until stiff peaks form. Fold in with crème patissiere to make crème légère.

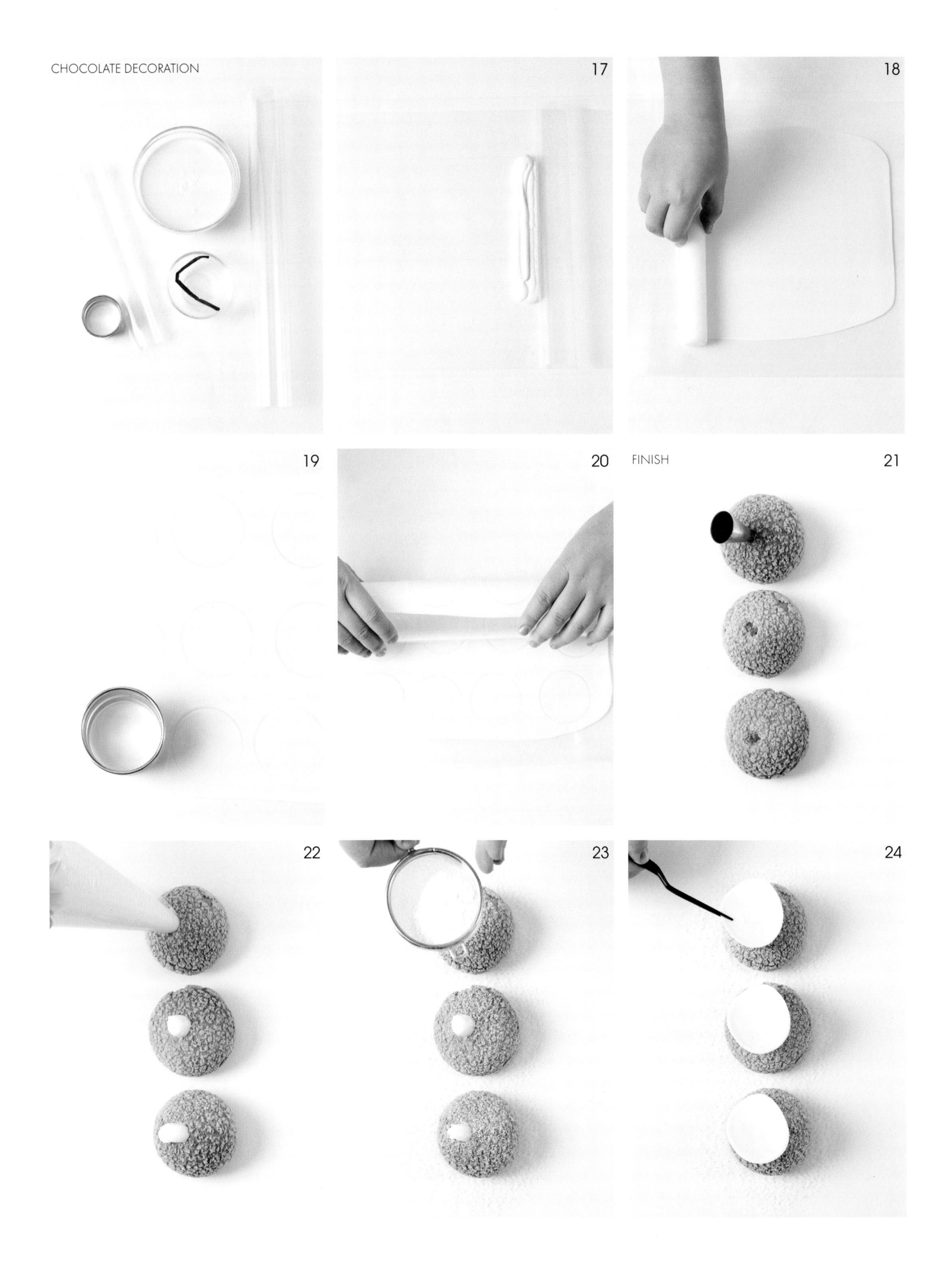
CHOCOLATE DECORATION
17
18
19
20
FINISH
21
22
23
24

초콜릿 장식물

17. 두 장의 투명 필름 사이에 바닐라빈을 섞어 템퍼링한 화이트초콜릿을 적당량 부어준다.
18. 필름을 덮고 밀대를 이용해 균일한 두께가 되도록 밀어편다.
19. 5.5cm 원형 커터로 커팅한다.
20. 아크릴 파이프에 감아준다.

마무리

21. 원형 깍지를 이용해 구워낸 슈에 크림 주입구를 만든다.
22. 크렘 레제를 가득 채운다.
23. 슈거파우더를 뿌린다.
24. 초콜릿 장식물을 올려 마무리한다.

CHOCOLATE DECORATION

17. Place a small amount of white chocolate tempered with vanilla bean seeds between two sheets of plastic film.
18. Roll out to even thickness using a rolling pin.
19. Cut using a 5.5cm diameter round cutter.
20. Wrap around an acrylic pipe.

FINISH

21. Using a small round piping tip, make three holes on the bottom of the baked éclairs.
22. Fill with crème légère.
23. Dust with powdered sugar.
24. Place the chocolate decoration to finish.

22 CHOUQUETTE

슈케트

ingredients

파트 아 슈

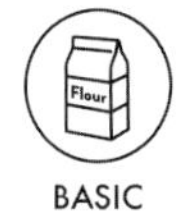

BASIC

GLUTEN FREE

하겔슈거(우박설탕) 200g

아몬드분태 200g

PATE A CHOUX

BASIC

GLUTEN FREE

200g Hagel sugar (Pearl sugar)

200g Almond, chopped

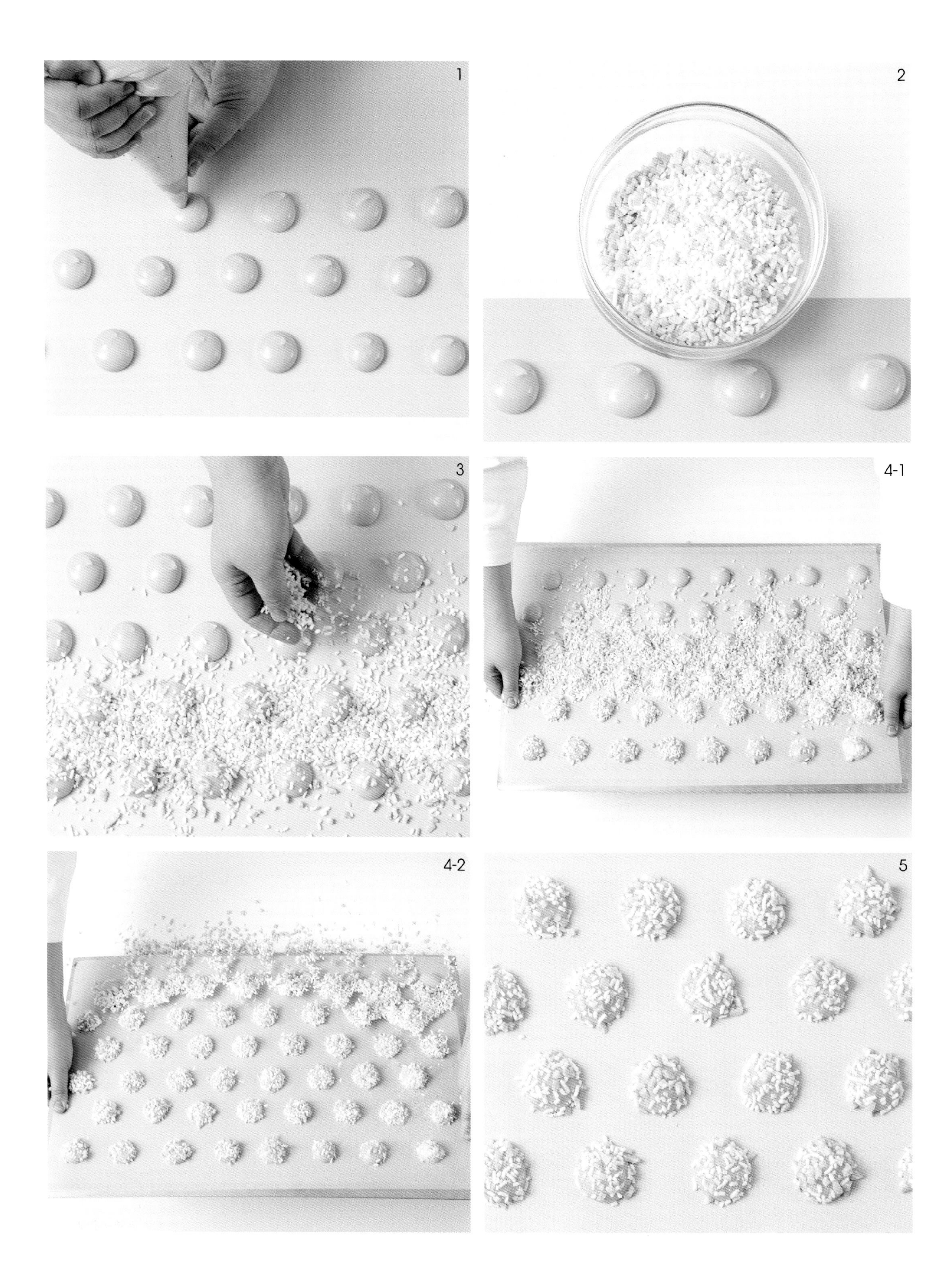
1
2
3
4-1
4-2
5

Process

1. 원형 깍지(지름 1cm)를 이용해 지름 3cm 원형으로 슈 반죽을 파이핑한다.
2. 하겔슈거(우박설탕)와 아몬드분태를 1:1 비율로 섞어 준비한다.
3. 슈 반죽에 **2**를 듬뿍 뿌려준다.
4. 테프론시트가 움직이지 않게 손으로 고정한 상태에서 철판을 움직여 하겔슈거와 아몬드분태가 슈 반죽에 골고루 묻도록 한다.
5. 160℃로 예열된 오븐에서 25분간 굽는다. (컨벡션 오븐 기준)

1. Using a 1cm round nozzle, pipe choux paste into 3cm rounds.
2. Mix 1-part hagel sugar (pearl sugar) with 1-part chopped almonds.
3. Sprinkle sugar mixture(**2**) generously on top of the piped choux.
4. To make sure the Teflon sheet doesn't move, hold it together with the tray and move it around to let the hagel sugar and chopped almonds evenly coat the choux.
5. Bake for 25 minutes in an oven preheated to 160°C (convection oven).

PATISSERIE
by
GARUHARU

TASTE

계절에 따른 가장 좋은 재료를 탐구합니다. 이렇게 찾은 재료의 맛이 디저트에 충분히 표현되었는지, 맛의 밸런스가 조화로운지 확인합니다.

We explore the best ingredient for each season. We make sure that the taste of the ingredients found is sufficiently expressed in the dessert and that the balance of the taste is harmonious.

TEXTURE

테크닉적으로 좋은 텍스처를 완성하는 것에 더해 하나의 디저트 안에서 단조롭지 않은 다양한 텍스처의 재미를 느낄 수 있도록 구성합니다.

In addition to mastering a technically good texture, we try to compose to experience the fun in various textures in one dessert that will not be monotonous.

DESIGN

디저트의 맛을 연상시킬 수 있는 포인트를 담은 간결한 디자인을 추구합니다.

We pursue a simple design with an emphasis that resembles the taste of the dessert.

Team GARUHARU

가루하루의 시작부터 지금까지 새로운 시도와 도전에 늘 열정적으로 동참해주는 가루하루 팀에게 고마운 마음을 전합니다.
성실하고 열정적인 재능 있는 동료들과 함께여서 새로운 시도를 주저하지 않고 무모했던 도전의 과정을 즐길 수 있었습니다.

I would like to express my gratitude to Team GARUHARU for their passionate participation in new attempts and challenges since the beginning of GARUHARU. Team GARUHARU for their passionate participation in new attempts and challenges. With these talented teammates who are sincere and enthusiastic, I was able to enjoy the process of reckless challenges without hesitating to try new things.

GARUHARU
SEOUL

GARUHARU MASTER BOOK SERIES 1

ÉCLAIR by GARUHARU

First edition published	May 20, 2020
Fifth edition published	February 20, 2023

Author	Yun Eunyoung
Translated by	Kim Eunice
Publisher	Han Joonhee
Published by	iCox Inc.

Plan & Edit	Bak Yunseon
Design	Kim Bora
Photographs	Park Sungyoung
Stylist	Lee Hwayoung
Sales/Marketing	Kim Namkwon, Cho Yonghoon, Moon Seongbin
Management support	Kim Hyoseon, Lee Jungmin

Address	122, Jomaru-ro 385beon-gil, Bucheon-si, Gyeonggi-do, Republic of Korea
Website	www.icoxpublish.com
Instagram	@thetable_book
E-mail	thetable_book@naver.com
Phone	82-32-674-5685
Registration date	July 9, 2015
Registration number	386-251002015000034
ISBN	979-11-6426-105-5

- THETABLE is a publishing brand that adds sensibility to daily life.
- THETABLE is waiting for readers' opinions and contributions on books. If you would like to publish a book, please send us a brief project proposal, your name and contact information to our e-mail address; thetable_book@naver.com.

- Misprinted books can be replaced at the bookstore where it was purchased.
- Price indicated on the back cover.